The Day Without Yesterday

Stuart Clark's career is devoted to presenting the complex world of astronomy to the public. He holds a first class degree and a PhD in astrophysics and is a Fellow of the Royal Astronomical Society. In 2000, the Independent placed him alongside Stephen Hawking and Professor Sir Martin Rees as one of the 'stars' of British astrophysics teaching.

He divides his time between writing books and his blog, *Across the Universe* for the *Guardian*. He is a consultant to *New Scientist*, writes features for *BBC Focus* and *BBC Sky at Night* and is a former editor of *Astronomy Now* magazine. Until 2012 he was senior editor for space science at the European Space Agency.

Praise for Stuart Clark's trilogy:

The Sky's Dark Labyrinth

'A vivid, thrilling portrayal of the lives and work of Kepler and Galileo . . . Books like this transform the way we access and understand our view of history'
Lovereading UK

'Sit under the stars and wonder, not just at their eternal beauty and mystery, but at the courage of the men who risked their lives so we could understand them'
Daily Mail

'Usually when reading a novel based on historical figures and true events, I find myself at some point asking what it's based on. It's a testament to Clark's ability to tap into the seventeenth-century mindset that this time the question never arose'
Historical Novel Society

'I could all but smell the streets and markets of seventeenth-century Prague in this novel'
Nature

The Sensorium of God

'Stuart Clark has once again invoked the great debates using enough science to help the reader's understanding without over-blinding laymen like me. The concluding volume will be about Einstein. My breath is bated'
Kathy Stevenson, *Daily Mail*

'Clark does a sterling job of covering the tricky period when scientists were the superstars of society'
New Scientist

'The best historical fiction goes beyond dates and events, giving historical figures emotions, achievements and failings. This is very much the case here, where petty squabbles sit beside philosophical debate in a rounded picture of great men and ideas'
We Love This Book

*****

BBC Sky at Night magazine

The Day Without Yesterday

Stuart Clark

THE SKY'S DARK LABYRINTH TRILOGY
Book III

Polygon

Paperback edition first published in Great Britain in 2014
by Polygon, an imprint of Birlinn Ltd

Birlinn Ltd
West Newington House
10 Newington Road
Edinburgh
EH9 1QS

www.polygonbooks.co.uk

Copyright © Stuart Clark, 2013

First published in 2013

The moral right of Stuart Clark to be identified as the author of this work
has been asserted by him in accordance with the Copyright, Designs and
Patents Act, 1988.

All rights reserved. No part of this publication may be reproduced, stored or
transmitted in any form, or by any means electronic, mechanical or
photocopying, recording or otherwise, without the express written
permission of the publisher.

ISBN 978 1 84697 282 9

British Library Cataloguing-in-Publication Data
A catalogue record for this book is available on request from the British
Library.

Typset by IDSUK (DataConnection) Ltd
Printed and bound by Clays Ltd, St Ives Plc

Content approved by NMSI Enterprises/Science Museum. Licence no. 0283.

PART I
Space

EDINBURGH LIBRARIES	
C0045887608	
Bertrams	09/04/2014
	£8.99
MS	DO NOT USE

1

Berlin, Germany
1914

Had the young men not been walking in rows, the physicist would have considered them a mob. Parading through the city in three-piece suits, fingers stained from their university inkwells, the youths held their straw boaters aloft and sang a rowdy version of 'Deutschland Über Alles'.

The march had brought the city to a standstill. Open-topped cars stood idling by the roadside, filling the air with fumes while the occupants craned to see. Bus carriages were similarly stalled, their horses nodding and snorting as passengers hung from the windows to applaud, forcing the students to raise their voices. Here and there, when a cheer was not enough, a clenched fist punched the sky.

The physicist had stumbled into the mayhem while lost in thought, hurrying through the streets. Snorting in disgust, he lowered his gaze and ploughed on, moving upstream. It was impossible not to bump shoulders; there must have been a hundred or more in the crowd, bold and boisterous. Emerging at the rear of the pack, he was surrounded by a straggle of older men and women egging on the lads.

He snorted again, louder this time. At least the boys had the excuse of youth.

'Albert! You're going the wrong way.'

The voice took Einstein by surprise. The lanky figure of Max Planck was standing on the nearby pavement, watching him through wire-rimmed spectacles. He grinned beneath a heavy, greying moustache. 'Beautiful day, isn't it?'

Einstein pursed his lips. 'I do not call *that* beautiful.' He jerked a thumb in the direction of the march.

Planck lifted his hat to run a handkerchief over the dome of his head. 'Heading back to the university?'

'No, I can't stop. I'm late already.'

'Nothing wrong, I hope?' Planck replaced his hat.

Einstein forced a name from his lips. 'Mileva.'

Planck's face fell. 'I wish you the best.'

Einstein nodded in acknowledgement and hastened away. He reached his destination soon after: a house in the expensive part of town, stone-built and three storeys high, it exuded power and achievement. The shiny black door opened before he reached the top step, robbing him of the chance to compose himself, and the entrance filled with Fritz Haber.

Like Planck, Haber was bald and bespectacled; it seemed to be the fashion for anyone over fifty at the Institute. Unlike Planck, Haber would have suited a uniform. He was strong, upright, eyes burning with self-assurance. There was a mild scowl on his round face.

'Sorry I'm late, Fritz. The streets are packed.'

'Another rally?' asked Haber, lifting his glasses to glance eagerly down the street.

Einstein nodded. 'They make it feel like a carnival.'

'Well, you can't keep a people down forever.'

Einstein swallowed a reply; now was not the time to debate imperialism. He looked over Haber's shoulder into the empty hallway. 'Is she here?'

'Mileva? Yes. Shall we begin?'

Einstein squared his shoulders, nodded tightly and followed Haber into the house. He was shown into the front room with its upholstered armchairs and cushions, wall hangings and paintings. A dresser displayed a collection of ornamental plates, each decorated with Teutonic hunting scenes, and a selection of freshly cut summer flowers filled the fireplace.

Einstein suppressed a twinge of envy, not for the furnishings but for the success they represented. Haber's nitrogen research had led to the manufacture of artificial fertiliser. He was a hero; his reputation and fortune were secure. He could please himself with his research these days. Einstein chased away the image of his own papers, unfinished on his desk at home, scored through with deletions and rewrites.

His host sank comfortably into one of the plush chairs and indicated that Einstein should do the same, but he could only bring himself to perch. He was dressed in his very best: an unseasonal dark suit, necktie and freshly starched winged collar. Such a get-up always made him feel awkward.

'Remember, Albert, I'm not your colleague today. I'm your friend.' Haber's mellow voice purred.

'I know that, Fritz.'

The chinking of china drew Einstein's attention. Haber's petite wife was carrying in a tray of tea.

'Clara. How are you?' said Einstein.

'I'm well. Thank you.' She poured, not meeting his gaze. Then she left the room in silence.

He took a sip of his drink. As he replaced the cup, it rattled against the saucer. 'Fritz, the situation has to end. You've been so generous to take in Mileva and the boys – I can't thank you enough – but they must leave.'

'Hush, Albert,' soothed Haber. 'The solution is agreed.'

Einstein nodded grimly. 'Then let us see that it is carried out quickly. They must return to Switzerland without further delay. Mileva knows this. We have grown too far apart. I am consumed with my work. This theorem inside me, it's alive – I can feel it, restless, kicking. It must be born or it will destroy me. Mileva knows that she is in its way as long as she remains.'

'Albert, listen to me. She does not need to leave. She has agreed to your terms.'

Einstein stared at Haber. 'She has agreed to leave my study without protest if I so request?'

Haber nodded.

'She will stop talking to me at once upon my say-so?'

Haber nodded again, his expression souring. 'And the rest, exactly as you have written.' He unfolded the sheet with Einstein's handwritten list and brought it into focus. 'She must clean and keep my study neat but not interfere with my papers. She must travel separately from me. She must launder my clothes and provide me with three meals a day. She must expect no intimacy, nor reproach me in any way. At my command she . . .'

'Enough, please.' Einstein turned his head one way then another in an attempt to conceal his embarrassment. Listening to the demands like this, he could hardly believe that he had written so callous a document. In the isolation of his study, his pen had provided the exorcism that his mind had needed, but to hear it back like this . . . He had demanded that his wife obey him like a dog. His skin prickled.

The boys had been tired and fractious the night he had written the list; the pounding of their feet around the apartment had been deafening. He had pulled open his study door and caught a glimpse of his wife in the corridor playing some sort of chasing game with them. 'Can you not keep them calm?'

Then Mileva had badgered him to join the meal or it would spoil.

'What could possibly spoil about your cooking?' he spat over his shoulder.

Mileva had set her hands on her hips and the argument had started. In the morning, Haber had offered to lodge her until they could sort out their troubles.

Einstein's gaze must have dropped during the reverie, because the next things he became aware of were his shoes, scuffed where he had forgotten to polish them and dusty from the walk across town. He shuffled his feet in a futile attempt to hide the shoes and looked up, straight into Mileva's eyes; she was watching him from the door.

Her hair had been freshly washed, and she wore a new cream dress, with puffed shoulders and a bow at the neck. It harked back to a style she used to wear. Perhaps it wasn't new; perhaps it was an old favourite from a time when they had been happy. Clara stood behind her like a shadow.

The two men stood up.

'You win, Albert,' said Mileva in a steady voice. 'I agree to your terms.'

'You don't understand. It would be a business arrangement, not a marriage. You would be little more than a servant.'

'Your letter makes it clear that the personal aspects of our relationship would be reduced to a tiny remnant.'

'How can you settle for that, Mileva?'

'Because . . .' She faltered, her eyes beginning to moisten.

'We must separate. We hurt each other too much.'

She looked as if he had just thrust a knife through her ribs. 'You refuse me even though I agree to everything you demand?'

He nodded mutely, his stomach a vortex. There could be no pretence any more.

Clara stepped closer. She placed her arm around Mileva's shoulders, which were beginning to shake, and led her away.

Einstein could feel Haber's gaze on his back.

'I'm sorry, Fritz, but it's for the best.' The room was stale and airless. 'I'll see myself out.'

When the train station's tall façade hove into view the next day, Einstein found it difficult to swallow, as though someone's hands were around his throat. His pace slowed the nearer he drew to the lofty windows and giant curved roof until the bell of an oncoming electric tram brought him back to life. He skipped off the tracks into the shade of a tree, where he loosened his tie and let his heart rate settle as much as it was going to that day.

People came and went from the terminus in droves. From the cigar-shaped Zeppelin passing high above they must have looked like ants streaming to and fro. He caught sight of a couple, arm in arm, and a mixture of hatred and longing built inside him. It was all he could do not to turn his back on the station and stride off down the road. He shook the thought from his mind and forced himself to march inside.

Mileva and the boys were waiting on the busy platform. Haber stood with them but stepped aside as Einstein approached.

Hans Albert was as tall as his mother's shoulder now. The ten-year-old's thick black hair was combed into a fringe that hid his forehead and accentuated his puzzled eyes. On Mileva's other side, less than half her height and clinging to her hand, was Eduard, little Tete. He still seemed a baby to Einstein, even though he had to be four now. He seldom spoke, mostly contenting himself with sucking his fingers and grinning.

'Your tie's undone, Albert,' said Mileva.

'Sorry.' He mumbled and repaired the knot. 'Where are your things?'

'You don't have to worry, we're going . . .'

'That's not what I meant.'

'The guard's loaded the trunk already.'

Mileva watched him from dark-circled eyes. Her hair was wiry, barely under control, not at all like it had been in Haber's apartment.

'When did you last sleep?'

She met his gaze with a challenging look. 'I don't think I've slept properly since you first visited Berlin.'

Einstein caught the weight of her answer as though it were a punch. She was referring to his trip two years ago to visit his mother, who had moved to Berlin from Italy after being widowed. At the time, Mileva had been unaware who else he would meet in the city. 'I knew you wouldn't like it here. I'm sorry,' he said hurriedly.

She locked eyes with him. 'Why are you here, Albert? In Berlin, I mean. Why did we leave Zurich for this? It can't be for the German people. You hate them and their pride, otherwise you'd have kept your German citizenship. Remember? You're Swiss now.'

Einstein tightened his jaw.

'Can't you see there's a war coming?' she continued, sarcasm gilding her delivery. 'You're going to be stuck here, with people you'll never agree with.'

'I still hope that common sense will prevail.'

'Don't be a fool. The parades, the marches; these people want war. They think they can just grab land from the French and no one will resist them. Switzerland's neutral. You were happy there.'

'They have offered me my own institute here, with staff.'

'Where is it, Albert? Not a word of it since we arrived, and who is going to build a new physics institute now? What good is science in wartime? Oh, you used to be so happy in Zurich. Marcel is still there.'

'Marcel?'

'Yes, Marcel.'

'Marcel isn't . . . isn't in the same league as the men here. Planck, Haber, even Nernst . . .'

'Same league? Listen to you. Marcel helped you recently because you couldn't do some piece of mathematics. What league are you in? Where's the Nobel prize you keep promising me? It's been nearly ten years since your paper on relative motion – it was going to make you world-famous, remember? We used to talk, discuss your work. You used to share. Now look at you. Your whole reputation is based on previous achievements.'

'But I'm so close now . . . The extension to relative motion is . . .' Again the image of his scattered papers assaulted him, sparking anger. 'Oh, why couldn't you have been more like Clara? That's the tragedy here. She knows how a wife should support her husband. He's a lucky man.'

Mileva glanced over at Haber, who was doing a very bad job of pretending not to eavesdrop. She fixed her granite expression on Einstein again. 'You just talk without thinking, don't you, Albert? Clara understands what being a wife is really all about? You men don't have to give up anything. Clara knows exactly how I feel.'

The train engine's whistle bit into the air. Einstein took the opportunity to break away, squatting in front of Eduard. Both sons had inherited his full lips and cleft chin. Hans Albert had his father's eyes, but Eduard's were more widely spaced and, at the moment, blank.

'Now, you be a good boy for your mother.'

Eduard nodded solemnly, sucking a finger of his free hand. Einstein briefly touched the little boy's shoulders, then straightened and turned to his elder son. After an awkward pause, Einstein stuck out his hand. Hans Albert stared at it before placing his own in it. 'Write to me often, son. Tell me everything you do.'

There was no strength in his son's grip and Einstein let it drop.

He turned back to Mileva, who was watching him contemptuously. 'I don't intend to ask you for a divorce.' Some small hope kindled in her eyes. 'All I require is that you stay in Switzerland with the boys.'

'But you need someone to take care of you.' A hint of pleading had crept into her voice.

He almost said *I have someone to take care of me* in order to terminate the conversation. Though he remained silent, his face must have given him away.

Mileva nodded, her suspicions confirmed. 'Give my regards to your cousin,' she said frostily.

'You mean Elsa?' He forced innocence into his voice.

'Who else?' she spat. 'You'll be round there before the train's even reached the suburbs. She's the real reason you want me gone.'

Einstein's eyes darted around. Hans Albert was staring at him with bleak attention. Eduard was lost in four-year-old thoughts, pawing idly at his mother's skirt.

'Goodbye, Mileva,' he said formally. 'Don't miss your train.'

The image of Mileva climbing on to the train filled his mind, her stiff posture and defiant shoulders as she ushered his sons aboard. Some paralysis had forced him to watch, to wait for the final venomous look she had launched at him just before the guard shut the carriage door.

He had thanked Haber for escorting her and then trudged from the station, clawing off his tie. His feet had begun to carry him towards Elsa's, but his wife's final accusation rang in his ears and he refused to give in to the knowledge that she had been right.

He had bought a coffee and sat on a wire-framed chair in a small café. Everywhere around him there was animated talk of war, the joy of conquest and the anticipation of destiny about to be fulfilled. He had lasted less than five minutes in there; eventually banging down the cup and slapping a few coins onto the table.

With nothing better to do, he had returned to his apartment, and as soon as the door shut behind him, he wept uncontrollably.

Now, as the sun began its slow dive towards the horizon, he began to perceive the stillness of which his home was capable. There was just the occasional muffled sound from the family in the apartment below, the whiff of onions or some other cooking. He felt strangely calm, drained.

There would be no interruptions now. Nor ever again.

His apprehension came from a new sensation growing inside him. It took him a puzzled moment to recognise it, but there could be no mistake. It was optimism: barefaced, naked optimism. Warm and comforting, it wrapped its arms around him. As it did so, mathematical symbols began to unfurl in his brain and Einstein reached for his pen and notepad.

It was days before Einstein emerged for anything more than a snatched meal or a trip to the post office. His first thought had been to visit Elsa, who was waiting for him across the city, puzzled by his absence. They had exchanged letters but he couldn't see her yet, not until the exorcism was complete and he had rid himself of the last vestige of Mileva.

Clara knows how I feel . . . His wife's parting words lingered in his thoughts like a thief in the shadows as he forced himself into a stiff suit and collar. He cursed every button but was determined to show that he was not helpless without a wife. He thought about visiting his mother but decided he could not face her gloating. She had been against Mileva from the start.

With a final tug of his waistcoat, he set off for the one place he knew he would feel at home.

The honey-coloured stonework of the university was a welcome sight. He slipped across the courtyard, watched only by the statues on the roof of the great portico, and scanned the drifting groups of students with suspicion. Which of these young men, now so earnestly carrying their books, had been in the rally the other day?

The long and noisy corridors were dappled by patchworks of sunlight from the windows. When he turned the final corner he saw Haber loitering, hands thrust into his pockets, trying to look nonchalant, thick lips pursed in thought.

'Fritz?'

Haber gave a sheepish smile and nodded towards the window at the end of the corridor. 'I saw you coming.'

'Are you well?'

'I am.'

'And Clara?'

'Also well.'

Einstein turned the key in the lock and showed him in. The office was a bit dusty but otherwise tidy, unlike his study at home. There were a few official university documents on the desk here that Einstein had yet to get round to completing. No matter.

Haber leaned against an empty bookcase as if he owned the place. 'Did Mileva arrive safely?'

Einstein set the chair's wheels squeaking as he sat down. He nodded; her letter had arrived yesterday.

'Good,' said Haber. 'Perhaps some time apart will . . .'

'Are you and Clara all right?' Einstein asked the question sharply.

It took Haber a moment before he smiled. 'Yes, perfectly.'

'Mileva wanted to be a physicist.'

'She told us that, while she was staying.'

'She failed her exams when she was pregnant with . . .' Einstein ran a hand through his tousled hair. This was not a direction he wanted to take. 'Clara doesn't regret trading her profession for marriage, does she?'

Haber looked thoughtful. 'I think she sometimes feels a little frustrated. She was a good chemist before we married. Why do you ask?'

'It's just . . .' Einstein shook his head, 'Nothing. Forgive me. I'm rambling. Few things on my mind, that's all.'

Haber drew near and placed a hand briefly on Einstein's shoulder. 'Of course. Are you working?'

'Indeed, a little slowly but making progress.'

'You still think you can fix the mathematics?'

Einstein turned to face him. 'I'll have the final formulation soon, and within the month Erwin will have the observations.'

'So you got the money together for the expedition?'

Einstein nodded. 'Erwin's in the Crimea now, should have reached the site a few days ago. He's probably all set up and ready, just waiting for the eclipse.'

Haber looked alarmed. 'The Crimea? Then he's in danger.'

'Why? The war – if it comes – will be to the west.'

Haber snorted. 'Where have you been? The newspapers are full of it. The Russians are mobilising. All that's needed now is a formal declaration of war.'

'That's impossible. Germany cannot fight a war on two fronts. They'll have to see sense and negotiate.'

'There can be no negotiating with the enemy.' Haber's pale eyes had turned to steel. He laughed, a little embarrassed. 'Cheer up. It's probably no loss. Erwin's a bit hasty from what I hear, prone to errors.'

'He's the only one who's shown any real support for my extension of relativity,' said Einstein indignantly.

'Oh, don't be so dramatic,' Haber purred, all calm restored. 'We all support you.'

'Yes, but he *believes* me.'

2

Feodosiya, Russian Empire

Despite the acceleration of his heartbeat and the sudden shortness of his breath, Erwin Freundlich was finding it easier than he would have imagined to stare down the barrel of a loaded rifle. It was aimed at the patch of skin between his eyes. At the other end of the barrel he met the dark eyes of his captor across the gunsight. Freundlich found himself thinking they were too feminine for a man, even one as young as this soldier.

'You are German, yes?' asked the smug-looking, grey-haired commanding officer at the young soldier's side. Squeezed into a uniform that once must have fitted, the officer exuded the air of a pampered child.

The astronomer glanced at his two travelling companions, similarly held at gunpoint, and nodded his head.

'Name?' The commander's accent was thick.

'Erwin Freundlich.' A spot of rain from the thickening clouds landed on his left cheek and rolled into his thick moustache. It felt like a spider crawling across his skin.

The commander stepped closer, and the astronomer saw clean over the navy-blue peaked hat with its gold piping. He could see the waters of the Black Sea beyond and hear the sound of the waves, softened to a soothing murmur by their distance. It mixed with the rustle of the breeze around the camp and the pumping of blood in his ears.

'You're all under arrest,' said the commander, chin lifted, eyes gleaming.

'Why? I have the correct permits. Everything has been . . .' He made to indicate the battery of cameras strapped to the sturdy wooden framework, pointing up at the sky, but stopped when the barrel of the gun jerked.

'No sudden movements, if you please,' warned the commander. 'Our countries are enemies now. Your government has declared war.'

Freundlich's mouth went dry.

The commander could barely keep the smile from his face as he finished his little speech. 'And that makes you our prisoners.'

They were forced to abandon the camp, leaving the cameras uncovered and the tarpaulins flapping, and marched to the police station at gunpoint, hands on their heads. Freundlich could feel the sweat running down his sides and the occasional raindrop along his jawline.

The townsfolk stared at them mostly with bemusement or confusion on their faces, but a few glared with hostility. Just the day before, the Germans had been moving around with anonymity; Zurhellen had lavished praise upon the local pastries while brushing the crumbs from his goatee, and Mechau had been watching the fishermen land their catches.

The police station itself was a damp stone affair just one road back from the harbour front. The three men were stripped of their papers and possessions – watches, pens, glasses – and taken deeper into the station.

'At least we're in the same cell,' said Mechau, laughing nervously as the door was shut solidly behind them and the key was turned.

Zurhellen stood glowering at the solid wooden barrier, fists flexing. 'How dare they? We're German citizens!'

'That's precisely why they've done it, Walther,' said Freundlich.

The comment drew a sharp glare. 'We can't just sit here and do nothing.'

'There's nothing else we can do.'

Freundlich studied the cell. There was a set of bunk beds lining both walls, a high window faced the door and there was a bucket in the corner. Clearly the room was not designed for lengthy habitation. Judging by the smell, it was probably where they locked up the drunks overnight.

'What are they going to do to us?' Mechau was sweating profusely, his hands shaking visibly.

The sight unnerved Freundlich. If the lad panicked, it would only make things worse. 'Nothing, nothing at all. They'll keep us here until they've verified that we're no threat, and then they'll release us and we'll go back to our work. We'll take the readings and travel straight home.'

Zurhellen sneered. 'We're at war! You can forget astronomy now.'

Mechau looked from man to man, lips quivering.

'Walther, please. I'm in charge of this expedition,' said Freundlich.

Zurhellen's eyes narrowed. He was the eldest here by a clear five years. Freundlich held his gaze, then turned to Mechau.

'We'll be released soon, don't you worry.'

Mechau let out the breath he had been holding. So did Zurhellen, but in an angry rush.

The commander looked over his spectacles at Freundlich. 'And you expect me to believe you? Light bending round the sun?'

Freundlich spread his hands in supplication, talking quickly. 'Have you never played in the rivers? Tried to catch fish with your bare hands? When you thrust your hands into the water, the fish is not where your eyes have told you it is.

Or a stick! A stick never looks straight when you place it in a glass of water. It seems to bend below the waterline because of refraction. Things aren't where we think they are. Gravity does the same. The sun will make the stars look as though they're in different places when it passes in front of them.'

'Those are some powerful-looking cameras you've got there.'

'They come from the searches for Vulcan.'

'Vulcan?' The commander cast his eyes around the grubby office. A uniformed policeman looked on from the corner. Two armed soldiers were stationed outside the door.

'Yes, the planet Mercury is off-course,' Freundlich gabbled, 'I mean it's not following Newton's law of gravity. Astronomers thought it was being pulled by the gravity of an undiscovered planet, Vulcan – but no one can find it . . .' He tailed off.

The commander lit a cigarette with slow deliberation. 'Planets, you say. One minute it's gravity, the next it's planets pulling each other about. Sounds to me like you need to get your story straight. You know what I think?' said the commander, not waiting for an answer. 'I think those cameras look powerful enough to take detailed photographs of the port.'

'We're not spies.' Freundlich wiped his brow. His anxiety must be making it look like he had something to hide. The realisation made him sweat even more. 'We're here to take pictures of the stars around the eclipsed sun and measure their positions. It's a crucial experiment. You must let us continue.'

If the commander's fat cheeks had risen any higher from his grin, he wouldn't have been able to see over them.

'We're going to miss it,' said Freundlich, standing on the frame of a lower bunk to look through the window bars.

'There'll be other chances,' said Mechau.

'When? It'll be years,' snapped Freundlich, immediately apologising when he saw the technician flinch.

'Beggar your theories.' Zurhellen spoke from the opposite corner of the cell, where he was squatting against the wall. 'We're at war. We should be out there now. Fighting for the Fatherland. Fighting these pigs!' He raised his voice for the last sentence.

Freundlich stepped from the bed. 'It's not our war. It's the Kaiser's. You'll only make things worse if you keep taunting them.'

'Speak for yourself,' Zurhellen sneered. 'How could someone like you understand, anyway?'

'Someone like me?'

'Well, you're not fully German, are you?'

'My grandmother was Jewish, if that's what you mean.'

'And your wife.'

'She's as German as you are.'

Zurhellen stood and took a step towards Freundlich. 'Not with those eyes and that hair.'

'She can be German *and* Jewish.'

The two men faced each other. Zurhellen's nostrils flared. Then, with a grim laugh, he turned and headed back towards his corner of the cell.

Freundlich turned too and smiled weakly at Mechau, who was watching with terrified eyes.

When the eclipse came, the darkness was complete. Freundlich peered out of the cell window and his mind filled with the work they should have been pursuing: loading the glass plates into the cameras, taking the exposures one by one, capturing the fleeting appearances of the stars in the brief minutes of the artificial night.

Einstein thought that strong gravitational fields were slightly different from Newton's predictions. He hadn't completed the maths yet, but it could explain the motion of Mercury and produce the light-bending effect, too. Freundlich's photographs were to have captured any deflection, allowing the theoretician to finish his work.

The astronomer felt a sharp sense of loss. He drew back from the glass and let his forehead rest against the cold windowsill.

They would not have seen anything anyway. It was pouring down out there.

Freundlich was frogmarched to the interrogation room a few days later. They had been given little but bread and water since their capture, and the sense of light-headedness from the hunger was beginning to feel normal. The fat commander was waiting for him, wreathed in the blue haze of cigarette smoke. 'You're to be moved to a camp in Odessa. It'll take a day or so to make the final arrangements.'

'You're not going to repatriate us?'

The commander spoke as if addressing a child. 'You're prisoners of war.'

'What about the cameras and the equipment? We must be allowed to take them with us. They're on loan.'

The commander blew out a long plume of cigarette smoke and laughed. 'Those cameras now belong to the Russian government. Oh, cheer up. You'll like it in Odessa, we've got plenty of other Germans there already. You'll feel at home.'

Freundlich's insides tightened. 'I'll go and tell the others.'

'Not so fast. It's just you and the young lad going to Odessa. We've been listening to your big-mouthed friend. He's going somewhere else. We've got a special camp for people like him.'

3

Berlin

Einstein checked his appearance in the entrance hall's mirror, took a deep breath and stepped into the lift. He wondered whether Elsa would hear the whir and clank of the machinery and be waiting with the door open. She would be smiling, he thought; it always drew attention to her eyes, which were by far her best feature.

When he arrived on the landing to see her door firmly closed, he lost some of the momentum that had carried him there that morning. He rolled his shoulders to loosen them, fiddled with the flowers he was carrying, and knocked with his free hand.

Elsa Löwenthal's plump face transformed into an uncertain smile when she saw him. She touched one side of her hair, lifting the curls into place. 'At last,' she whispered before recovering her voice. The intonation of her Swabian accent imprinted its own stamp on his name. 'Come in, Albertle.'

'I've written,' he said, offering the flowers.

'From all that way across town, I know. I've written back, or haven't you noticed? Of course you have, but letters are not the same as being together, are they? Oh, aren't these flowers beautiful? Now go through and I'll put these in water.'

Einstein hesitated as she closed the door behind him. He should tell her now, get it over with, but the familiar smell of beeswax ambushed his determination.

'Well, go through,' she said, shooing him along. 'You know we don't stand on ceremony here.'

She disappeared past a sliding door into the kitchen and began to rattle around in the cupboards.

He placed his hat on the coatstand and went into the living-room. The windows were open, allowing the summer heat to escape and the sounds of the city to filter in. Even so, it was too warm. He stopped himself from clawing at his shirt collar and distracted himself by comparing the state of Elsa's apartment with what he was used to. There were no half-read magazines or dirty crockery cluttering up the place. On the sideboard was a single, silver-framed photograph of her late husband. It was a bluff really, she had divorced him four years before his death, but it was more convenient to pass herself off as a widow than a divorcée.

She swept into the room, summer-weight skirt swinging around her sturdy calves, carrying the flowers in a tall glass vase. He noticed that her skirt fabric was printed with red roses too. She positioned the vase on the sideboard so that the light caught the glass and danced through the water, then she made a final small adjustment to rotate it into just the correct viewing position.

'There, don't they look pretty? Brighten the place up.'

'I'm glad you like them, Elsa.' He fought to say more but could not coax words. It was a confounded sensation; when had he ever felt anything other than comfortable in Elsa's presence?

'Was it terribly difficult at the station?' she asked. 'Your letters made it sound awful.'

'It was for the best.' He looked around. 'Are Ilse and Margot here?'

Elsa shook her head. 'They're out watching the soldiers.'

'You let them go?'

'Now, now, Albertle. Let them have some fun. They'll stay together.' She gestured for him to sit and they took positions on either side of the fireplace. She sat forward, leaning towards him, knees properly together. 'I've been thinking. You could put your desk in the dining-room, near the window. We could move the plates off the dresser to give you some bookshelves. And you won't be bothered much, only at mealtimes, and then you'll be eating with us. Also . . .'

'Elsa, I'm not moving in with you.' There, he had said it.

Her brow creased and a puzzled expression crossed her face. Abruptly she rubbed her hands together. 'I'll get some tea. You must be parched from walking over here in this heat.'

Her departure raised a breeze in the room. After a few moments he followed, hesitating in the doorway to the kitchen, not daring to cross the threshold. He watched her draw water from a juddering tap.

'Get the cups down for me, Albertle. Make yourself useful,' she said over her shoulder.

He stood his ground. 'Elsa, it's not because I don't want to be with you. I just find that I . . . I . . . I know you're here for me when I need you. I can content myself with that for the time being. I'll see you often; we'll have all the trappings of being together but none of the getting under each other's feet. It'll be ideal, don't you think?'

'But how will it look when we're married? If we're not living together, I mean?' She swung the kettle to the stovetop.

'Elsa, I'm not asking Mileva for a divorce, just a separation.'

She stopped what she was doing and stared at him briefly. 'I'll get the cups then.'

'Elsa!' He hated it when she pretended not to hear him.

She froze, her face a mask.

'Elsa . . .'

Knocking like rifle-shots at the front door interrupted them. Einstein bit his lip in exasperation. Elsa was already in motion again.

'I'll get that. You get the cups.' She brushed past.

He reached up to massage his eyes but paused when he heard his name.

'Is Herr Einstein here? It's really most urgent that I see him.' The voice was female, young and unfamiliar.

Elsa's voice was full of suspicion as she tried to fend her off.

'Please, I have to see him,' begged the intruder.

Einstein stepped into the hallway and the young face flooded with relief. It was too round and her lips too full to be beautiful, but her nose was neat and her eyes shone chocolate-brown. Einstein knew they had met, but guiltily could not place her.

She rushed forward past Elsa, checking herself just before colliding with him. 'Erwin's been captured. You've got to get him back.'

He stared at the young woman in front of him.

'Herr Haber at the university told me you'd be here.'

'Do you two know each other?' Elsa asked from the doorway.

Einstein nodded, realisation flooding through him. 'Elsa, this is Frau Freundlich, Erwin's wife. We met last year, on their honeymoon.' He smiled at the visitor. 'But I didn't pay you as much attention as politeness deserved on that occasion, Frau Freundlich. I apologise.'

'Please,' she said, 'call me Käte.'

Their meeting had been in Zurich. Einstein had insisted – almost demanded – that the newlyweds detour for a day during their Swiss honeymoon. From the moment of their meeting on the station platform, the two men had talked non-stop about relative motion and its potential for understanding gravity. Käte had trailed in polite silence, contemplating the buildings and the scenery.

'I dare say that was his favourite day of the honeymoon,' said Käte, after Einstein had recounted the story to Elsa.

The older woman nodded with resignation and closed the front door. 'Take her through, Albertle. I'll get the tea.'

Käte stared into her cup, raising it occasionally but barely wetting her lips. Elsa made a few attempts at small talk but the conversation died quickly every time.

'At least he's alive.' Einstein handed back the typed letter informing Käte of her husband's capture. He had read it through twice before the words sank in.

'What can we do to get him back?'

Einstein felt as helpless as she did. Nevertheless, he spoke off the top of his head. 'Krupp in Essen helped sponsor the expedition. They must have some leverage with the government, they're a big firm.'

She nodded hopefully.

Einstein thought about the time it would take to arrange the meeting. The girl obviously needed comfort now. 'And we can go to the university and speak to Max,' he said.

'Herr Planck? Erwin has talked about him.'

'He's a good man. He'll know what to do – he backed the expedition, persuaded the Academy to provide money for the equipment. They'll be eager to bring this to a safe conclusion.'

Plan formulated, he ushered Käte into the hallway.

They were halfway to the staircase when he heard Elsa calling him from the apartment. He swung back to see her chasing after them. She was carrying his hat.

'You forgot this.'

'Thank you.' He lifted it to his head and went to turn away.

'Albertle, I lo—' The words died on her lips.

He looked into her eyes from under the trilby's brim. 'I do too,' he said.

The streets were full of young men again, only this time the straw boaters had been replaced by spiked helmets; three-piece suits had given way to grey uniforms and boots had substituted for shoes. Each man carried a rifle. Sheathed bayonets swung from their belts, and their marching feet pounded out a steady rhythm over which the cheering of the onlookers became a single continuous melody.

Unable to contain herself, a young woman leapt from the pavement and pinned a flower to a soldier's lapel. He broke rank to accept the talisman. She hugged him clumsily and rushed back to the pavement, face glowing red.

Einstein stared. At least the girl was neither Ilse nor Margot.

'Madness,' he muttered and guided Käte on.

It was quiet now at the university, the rush to enlist having drained the place of young blood. Long lines were stretching down the streets from the recruitment offices. When night fell and the doors were closed, pride dictated that those still waiting would stand their ground, sinking to the pavement to sing songs and slumber against the sun-warmed flagstones until dawn saw the offices open again.

'How can we lose with determination like that?' Einstein had heard one old woman say, clutching her bag of groceries.

He shook the memory from his head and led Käte into the university.

He saw Nernst first, and had to blink to convince himself that he was not hallucinating. The short, pot-bellied chemist was swaggering along the corridor in full army uniform. He waved a pair of riding gloves at Einstein. 'I'm a courier on the western front.'

Haber was with him, looking dapper as usual, fortunately still in civvies and wearing a faintly bemused look on his face.

Einstein tried to ignore the Fatherland's newest recruit. 'Where's Max?' he asked Haber.

'In his office.'

Nernst stepped into his path. 'This is our chance to prove ourselves. All three of us . . .' He noticed Käte and studied her for a moment. 'I dare say all four of us. We Jews can show our patriotism.'

Einstein looked down at him. 'By killing foreigners?'

'If it secures us an equal place in Germany. I'm sick of the suspicion against us.'

'Albert, you probably haven't been here long enough yet to notice it,' said Haber. 'We're tolerated one minute, ignored the next, then ridiculed, but never really accepted.'

'And you forget, I'm not German. I'll always be an outsider.'

'That makes being a Jew even worse.'

There was the crash of double doors rocking on their hinges. The four of them turned to see Planck striding towards them, his face flushed. 'Belgium is calling up its forces. They're going to fight!' he announced.

'But they're neutral,' said Haber.

Planck shook his head. 'Not any more.'

'Stupid fools,' said Haber. 'Why not stand aside and let us march to France? They'd have been under German protection, the war could have passed them by.'

Käte looked askance at Einstein, who tried to reassure her with a faint smile.

'Belgium resisting,' growled Nernst.

'Civilians have been sniping against our boys, and now it's a full mobilisation,' said Planck.

'They will pay for this outrage,' said Nernst.

Einstein's mouth dropped. 'Walther, that's enough of this nonsense.'

'They were neutral.' Nernst waved the fist carrying the gloves. 'They should have let our forces pass through, as is our right. This is . . . treachery!'

Haber closed his eyes and dropped his head; he seemed to be muttering something to himself.

'Fritz, please talk some sense here,' said Einstein.

Haber nodded slowly. 'Something must be done to end this war quickly.'

'The Belgians have no idea what they have brought upon themselves!' Nernst was almost shouting, holding a finger in front of his face like an exclamation mark. 'They stand in the path of a steamroller. They will be utterly destroyed. Gentlemen, I must go, I have duties to attend to.' With that, he turned and waddled away, swinging his arms, huffing and puffing.

In the silence that followed, Einstein took his chance. 'Max, I must talk to you.'

Planck's office was plush, with a wide desk that the physicist kept in an orderly fashion. There were three piles of paper, an almost spotless blotter and a shiny inkpot. Shoulder-height bookcases lined the walls, topped with vacuum tubes, flasks, clamps and other apparatus that looked somehow arranged as art, rather than stored in readiness.

From behind his desk, Planck tapped a long finger against his lips. Approaching sixty, he was a hangover from the nineteenth century. Although he had never sported lamb-chop sideburns to Einstein's knowledge, the long moustache was distinctly old-fashioned. Yet it suited Planck, conferring a patrician air.

He looked at Käte, his eyes as round as an owl's. 'Let's start with the consulate. See what they can do.' He lifted the receiver and placed his request with the operator.

Einstein exchanged hopeful glances with Käte as Planck was alternately connected and then passed back to the operator, talking to one person after another, his voice always patient as he sought the right official. 'Erwin Freundlich, an astronomer leading an eclipse expedition,' he said numerous times. Eventually, his face set like stone and he replaced the receiver a little too heavily in its cradle.

Käte's face drained of colour.

'They say they'll do their best to locate him. I'm sorry. They wouldn't promise any more information.'

Einstein could feel Käte staring at him.

'What if they shoot him? I've heard the Russians are capable of anything. His death will be on your conscience, Herr Einstein.'

Einstein kept his face impassive. He had thought of nothing else since Käte had arrived with the news an hour or so before.

4

Odessa, Russian Empire

The camp was a series of single-storey wooden huts, each one raised up on stilts to provide some protection from rats and a convenient place for the children to play.

The inhabitants were mostly expatriate Germans, with a smattering of Austrians. Many had been local inhabitants, rounded up as soon the hostilities had begun, and their conversations generally turned to expressions of betrayal by neighbours. Even if the war ended tomorrow, Freundlich suspected, it would take a herculean effort for life to return to normal.

As far as he could tell, no one had yet established himself as *de facto* leader of the refugees, and there were new arrivals each day to keep the population in constant flux. One morning he counted every spare bed, and then averaged the number of new arrivals over the week. At this rate the camp would be full in a little over a month.

He and Mechau had secured reasonable bunks in a hut far from the dusty exercise patch. They were away from the door but near a window. The building was so new it smelled of sawdust and the bedposts offered a ready supply of splinters.

'What are we going to do?' hissed Mechau one afternoon from the cool of the lower bunk.

'Nothing. We keep our heads down, do what we're told and everything will be fine.' Freundlich was lying on the thin mattress above. 'If they'd planned to shoot us they would have done so by now.'

His tousle-haired companion's head shot up beside the bed.

'Sorry, Robert. Bad joke.'

The young technician tried to laugh it off. It was an unconvincing sound. 'How long do you think we will be here?'

Freundlich looked at him squarely. 'I don't know. No one does.'

It might not have been a satisfactory answer, but it was an honest one.

After catching his fingers once too often on the bedpost, Mechau palmed one of the blunt knives from the mess-hut and used it to scrape the bunk's uprights. It proved such a laborious job that he took to scouring the compound for flints, and gradually the bedposts took on a smoother finish. He even talked about putting a chamfer on them.

Anything, thought Freundlich, *if it keeps him busy and stops him brooding.*

At night, Freundlich would point out the constellations through the hut windows. It was hardly ideal, but with the curfew in place it was the best that they could do. He was hoping for a final sighting of Antares before the ruby star slipped below the horizon to visit the southern skies for autumn and winter.

One evening as twilight was painting the sky, Freundlich led Mechau to the southern perimeter fence. They could see down the tumbling hill, over the town to the sea.

'There!' said Freundlich, his arm ramrod straight.

Mechau's small gasp told him that he had seen it, too: the twinkling point of blood-red light, just visible in the gathering violet of the night-bound sky.

'We don't even know what keeps them burning,' said Freundlich, 'yet they've shone for aeons.'

Mechau stared, captivated. 'How long have you been interested in the stars?'

'Always. I think they're the most beautiful things in nature: more so than mountains or trees or flowers.'

'More so than girls?'

Freundlich laughed, the first spontaneous laugh he had enjoyed since their arrest. 'You're right. Nothing could be that beautiful,' he said, thinking of Käte.

The bell rang out, breaking the spell. He tutted loudly and turned to glare at the tower.

'Come along, Erwin. Remember, we're keeping our heads down.'

In general the camp was a quiet place, which was why the sound of a woman's anguished voice outside their hut one afternoon immediately drew attention. Through the window, Freundlich saw a hunched grandmother fighting a tug-of-war with a boy over a length of sausage.

'Hey!' He jumped from the bed and bounded for the door.

Mechau rushed after him.

As he rounded the outside of the hut the boy broke away and ran, carrying his prize. The old woman began crying out at the top of her voice. 'It's mine. I've been saving it.'

'Come back here, you!' Freundlich gave chase but the boy's short legs were a blur. He was about to round the corner when another figure appeared. The boy slammed straight into the rotund chest, midway between the braces that held up heavy brown trousers.

'What do we have here then?' The man's accent was rough. The boy held up the food. Freundlich came to a halt as the man looked up with crooked eyes. 'Chasing a young-ster for food?'

The man lifted the boy and swung him round behind him. He was upon Freundlich before the astronomer could utter a word. Blow after blow rained down and then Mechau was in there too, hammering away at the assailant.

'Stop what you're doing at once!'

In his panic, Freundlich thought it was Mechau shouting. But the voice had the wrong accent. Then he heard the click of rifles being cocked.

Next morning every single one of the camp inmates was called to muster on the strip of land in front of the huts. The commandant was a surly creature who stood stiffly in front of them, flanked by more guards than Freundlich had realised patrolled the camp. Not even the commander's sharp uniform could make up for the fact that he was ugly. A giant black mole squatted on his left cheek.

'There will be discipline in this camp. I will not tolerate fighting over food or any black market in goods. You are fed and you are cared for. Can't you see that this is the best place for you in this war? You're safe here.'

Freundlich's jaw still smarted, and Mechau's right eye was swollen, from the blows that the bully had landed. The guards had been rough in pulling them apart, but had insisted upon hearing all sides of the story before providing witch hazel and rags to treat their bruises.

On and on barked the commandant. His lecture consisted of the same basic information delivered repeatedly and more forcefully with each iteration. Next he began to list the petty crimes that had occurred in the camp: insubordination, curfew-breaking, fighting over food . . .

Freundlich's attention returned to his smarting jaw until there was a stirring in the ranks around him. People were fidgeting nervously. Had he missed something? The guards were preparing for something and watching the ranks as if they expected trouble.

'I repeat,' said the commandant, 'if your name is called, step forward.'

So this was it, thought Freundlich. Punishment.

The list began; his own name was close to the top. He knew the guards had not believed him when he explained why he was chasing the boy. The old woman had disappeared, sensibly deciding that a lost sausage was preferable to a tangle with authority.

He trudged over to join the group of miscreants. One was a young girl clutching a teddy bear. What could she possibly have done? Loathing for the guards spiked within him.

'Robert Mechau,' thundered the commandant.

A cold hand gripped Freundlich's heart and he looked back at the ranks. His young companion was shaking so violently he was having trouble walking. Mechau stumbled and fell. Guards rushed forward and hauled him from the ground. Freundlich took him from them. Soon they were herded away at gunpoint.

As he walked, Freundlich's mind flooded with images of Käte. He had promised her so much. They had only been married a little over a year. What he hated most about this pointless ending was that he would not be able to apologise to her for messing it all up.

5

Diksmuide, Belgium

The German advance into Belgium was swift and utterly brutal. It cut through the land; it cut through the people. It forced the exhausted defenders back towards the French border and then came at them again, determined to slice through to Paris.

But, somehow, the Belgian line was holding, or at least it had been.

The columns of marching men were glimpsed first by the lookout on the church tower. Soon, their advance was being watched by the entire garrison from the defensive line in front of the town.

The new arrivals followed the road over the flatlands. The poplars that lined their way had been green when Georges Lemaître arrived there a month ago. Now they were golden, their leaves just clinging to the branches. There was at least some comfort in that. In Lemaître's twenty years, he could remember times when the lowland trees had been stripped bare overnight and the slide into autumn had been overtaken by a plunge into winter. He prayed to God that this would not be one of those years.

He was lying on the ground behind an earthen mound, rifle trained straight ahead.

'Reinforcements,' said the soldier next to him, revealing a toothy smile.

The approaching soldiers were indeed wearing Belgian uniforms, but Lemaître could see the dirt on their faces and the weary way they swung their arms. 'They're not reinforcements.'

The sentries lifted away the wooden poles that served as roadblocks and stood rigidly to attention as the fighters passed. The wounded came into view; some walking, with dirty bandages on their arms, legs and heads; others carried on stretchers. One man was slung over a horse, the officer atop the beast resting a hand on his back.

It could mean only one thing: Antwerp had fallen.

The realisation spread through the garrison like an early-morning frost.

'Just us left,' whispered Lemaître.

Later that day, he prowled the defeated army's makeshift camp. The soldiers were huddled around campfires, blankets around their shoulders, united by the smell of simple food and wood smoke.

Lemaître scanned every face, his frustration growing. A man with one arm in a sling was spooning food from a tin plate.

'Do you know Jacques Lemaître?'

With a curt shake of his head, the wounded man looked down at his meal and scooped another mouthful.

The dusk was gathering, but the thought of spending another night without having found his brother chilled Lemaître more than the autumnal air.

'I know him,' said a soldier. 'He was stationed at the bridge, rigging explosives. I saw him as I crossed ... The Germans knew we were planning to blow the bridge and were determined to take it.' The man bit his chapped lower lip. 'It was bad back there ...'

Lemaître turned away before the man could say any more and stumbled on. Not watching his step, he let his boots became entangled in some guy-ropes. Beside him, men cursed as their tent collapsed to the ground.

'Sorry. I'm looking for my brother.'

A hand came into focus. 'Let me help you.'

'Jacques,' he breathed.

The face was lined and weary but the boyishness remained.

They embraced roughly, slapping each other hard across the back. When they released each other, Jacques looked him up and down. 'Nice clean uniform.' His own was encrusted with mud.

Lemaître nodded, brushing off the dust from his tumble. 'Thank God you're alive. I can only imagine how it was in Antwerp.'

'A bit . . . difficult at times.' Jacques tried to flash a grin but it twisted horribly. He ended up looking away into the distance. 'They'll be coming after us, you know.'

Lemaître nodded.

'Ypres is still free, but it's here that the Germans really want.'

'I know, Jacques.'

Diksmuide was all that stood between the Germans and Nieuwpoort, the only Belgian port still unconquered. If the Kaiser's forces reached the coast, they would be able to slice westwards into France, outflank Ypres and swallow the remains of Belgium whole. Fear knotted itself around Georges's core. 'We don't need to talk about this now,' he said.

'I can't believe we've lost everything in less than two months.' His brother sounded angry.

'I don't think the French thought we'd even last a week. Plus you're forgetting something: it's not over yet.'

'You think we can defend this stretch?' Jacques's voice held a touch of sarcasm.

'Why not?'

'You don't know what it's like. You can't think for all the noise and the chaos; mud flying; smoke in your eyes. They're inhuman. They pound you with artillery night after night – can't sleep a wink – then they come marching towards you and nothing can stop them.' He took a great pull on the cold air, then spoke in a faltering voice. 'It's all so terribly noisy. My ears ring all the time now. Sometimes, when it's quiet, I think I'll go mad from them screaming at me, inside my head.' He regained his composure. 'Got any cigarettes?'

Georges pulled a crumpled packet from his breast pocket and handed it over. 'Bit battered, I'm afraid. I try to keep them for weekends. But you can have them.'

'How do you even know what day of the week it is?'

Georges did not like to mention the boring monotony of his days so far. It was all so regimented here that, during his time off, he had even made a fair amount of progress through the textbooks he had brought with him.

Jacques lit the cigarette in a single fluid movement that took Georges by surprise. The last time they had smoked together, his younger brother had been a clumsy beginner trying to look suave.

Papa used to tell them off when they came home smelling of tobacco, though he was hardly ever without a stub in the corner of his mouth himself.

'Do you think about Mama and Papa?' asked Georges.

'All the time.'

'Me, too.'

'It can't be easy for them. They're not young.'

Georges imagined the familiar Brussels streets brimming with German soldiers, his parents trying to conduct some sort of normal life around them.

'All these weeks I've been thinking of things to say to you, and now we're together I can't think of a single one.' Jacques took another deep drag on the cigarette.

'Then let's not talk.'

Jacques smiled, weariness filling his face. 'Sounds good to me.'

They squatted on the packed mud and watched the very last vestige of the day trickle from the land. There was a bloodstained sunset and, when that too disappeared, it left only the aromatic comfort of the tobacco smoke.

The German forces attacked at dawn two days later, not long after Georges and his patrol had arrived in one of the dykes in front of the town. At first he thought the sound was thunder, but there were no clouds in the sky. There were, however, silhouettes on the horizon, jagged outlines where the day before there had only been straight, flat land. Puffs of light erupted from the silhouettes and the roar of the artillery rolled across the plain shortly after.

Unexpectedly, the fear in his stomach subsided. At least the waiting was over.

The first explosions reached his ears as the shells crashed to earth and the air filled with the smell of damp grass. From the fountains of dirt off to his left, he saw that the rounds were falling short of their mark.

Another volley split the air and the field erupted again, closer this time, producing some nervous shuffling on the duckboards around him. 'Don't worry lads, nothing dangerous yet. Stay calm,' called the sergeant, an avuncular man who had been a butcher before the war.

Lemaître tried to picture him in a blue-and-white striped apron, chatting to housewives and slicing ham. Another volley of artillery fire brought him crashing back to the present.

He was part of a volunteer unit with only basic training. Clutching his spindly rifle as the shells ploughed up the field, Lemaître realised just how outclassed they were by the German guns.

The Belgians had some heavy weapons – they were barking from behind the dykes in response to the attack – but the shells were falling well short of the enemy lines. The

German bombardment was edging closer and closer to its mark.

Noise and confusion engulfed him. Stones and dirt were flying through the air, thudding off his shoulders and tin helmet. A second almighty blast came and he was blown backwards into the earthen wall. He lost his footing and found himself on the duckboards, pushing his glasses back into place. Around him, his companions were in similar disarray.

A third muffled explosion shook the trench. This time he saw a column of earth rocket into the air, a maelstrom of dirt in which he could just make out larger, whirling patterns. With sudden comprehension, Lemaître recognised the shapes as men being blasted off the ground.

The soldier next to him began to scream. Lemaître jerked round and made a quick inspection through mud-spattered glasses.

'You're fine! Not a scratch on you.'

Still the soldier wailed.

A fist shot past Lemaître and struck the frightened man on the cheek. His eyes widened in surprise and the moaning stopped.

'He said, you're fine,' said the sergeant, withdrawing his clenched hand.

German shellfire raked the trenches back and forth all day and all night. Under the cover of darkness the invaders crept closer, and by the next morning Lemaître could see individual soldiers moving about the enemy ranks if he squinted hard enough. They stayed just beyond the reach of the Belgian artillery.

The German guns fired, a rattling cacophony that shattered the air and sent Lemaître and his colleagues diving for cover. The ground erupted around them, blow after blow, explosion after explosion, each concussion sending shock-waves through the air to batter their eardrums.

As the last of the dirt fell back to earth, the ground began to quake unlike anything Lemaître had experienced before. He straightened his helmet and peered across the field.

Dear Lord have mercy . . .

The sound was coming from a thousand pairs of enemy feet charging towards them.

'All right, lads. Wait for it,' barked the sergeant.

Lemaître levelled his gun at the rushing horde. A quick glance left and right told him that his comrades were doing the same.

They waited as the enemy charged unchecked.

'For king and country! Fire!' yelled the sergeant.

Lemaître twitched his finger and felt the thump of the recoil. The ignition of his weapon's gunpowder conjured the smell of fireworks and utterly misplaced memories of family laughter and hot beer.

The Belgian artillery fired too, blasting small pockets of the attackers into the air.

Lemaître fired again, but the shakes had started. As he squeezed the trigger, the gun jumped skywards.

'We're not shooting pigeons!' the sergeant shouted.

Lemaître reloaded, choked down his fear and re-sighted his weapon. This time the enemy were so close he could see round pink faces under the spiked helmets, and he found it harder to pull the trigger. With the piercing crack of rifles, the shouts of the advancing troops and the drumming of their feet, a sense of unreality was growing inside him, as if he were a spectator to this surreal exchange.

Defensive fire from a machine-gun off to his right jarred him back to his senses. He planted the rifle-butt in his shoulder and squeezed the trigger. A figure fell from the advance, but there was no way for Lemaître to know whether it was his bullet or another that had found its mark. There were men dying all along that breaking wave now. Yet still their combined mass bore down on the town.

The machine-gun sounded again.

There was movement that Lemaître could not comprehend. He squinted for a better look, and with a desperate recognition saw that invaders were being ripped apart every time the machine- gun chattered. They could just have been uniforms cut to shreds, except for the red wetness that exploded out of them.

Lemaître's insides heaved. He spat foul-tasting bile from his mouth and angrily blinked back tears.

The sergeant's hand rested on his shoulder for just a second. 'No time for that, Georges,' he said quietly, then returned his voice to full volume. 'Ready, bayonets!'

But before Lemaître could fumble his bayonet into place, the attackers began to disappear from view, seemingly sucked down into the very earth itself. They were taking cover in the craters that their own artillery had dug.

'Grenades!' warned the sergeant.

Sure enough, dark shapes arced through the air. The pop of the detonations was strangely comical next to the thump of the big guns, but the effect was just as devastating.

Time became meaningless. Everything blurred into a continuum of gunshots and men screaming. Lemaître became an automaton, shooting by reflex every time he saw a shape appear over a crater rim or his sergeant called for a defensive volley so that the Belgian grenadiers could lob their own deadly packages. There was no thinking involved any more, just reaction.

They targeted the German defensive positions one by one. The machine-guns ploughed up the field, cutting lines across the hiding places, scything down anyone unfortunate enough to choose that moment to look up. The Belgian grenadiers hoisted small explosives in the same direction, and one by one the enemy nests were destroyed.

The shrill sound of whistles cut through the din, and with a final furious barrage of German grenades, the enemy began to retreat. Some men around Lemaître began to scramble up the dyke wall.

'Stick to the line, lads. Stick to the line,' shouted the sergeant. 'Don't follow them. Let them run.'

Lemaître watched the Germans along his gunsights until they became too distant to see as individuals. He could not bring himself to fire during the retreat – not into a man's back.

He dipped his head and pushed back the brim of his helmet. He felt sick, but the feeling was soon displaced by trembling exultation; every part of his being sparked with energy and he wanted to shout in triumph.

Ashamed of his reaction, he was glad to be detailed to help remove the bodies. He dragged corpse after corpse to the tumbrils, each one feeling a little heavier than the last. He tried not to look, but it was impossible not to recognise some of his friends. His exultation turned to horror and, with frightening speed, threatened to overwhelm him. He found himself gasping for air and ran a dirty cuff across his nose. The scratch of the wool distracted him from the urge to run.

Rubbing his hands angrily together, he reached down and grabbed the closest pair of blood spattered lapels. They belonged to a middle-aged man with an immaculately styled moustache. 'Come along, you lucky sod. Look on the bright side, the war's over for you.' He heaved the body on to his shoulders.

Around him, the German bombardment resumed.

6
Berlin

'You must think I'm very silly, Herr Einstein, insisting you come with me today.' Käte regarded him with an uncertain expression.

'Not at all,' he said, concealing his discomfort at being back in the train station. Perhaps the memory of Mileva and the boys was rendered more acute today because a letter had arrived, carrying another unfamiliar return address. His wife had become nomadic since her return to Switzerland, flitting from one friend to another.

Käte wore a long black overcoat and held her handbag tightly with both hands in front of her. She sported a wide-brimmed hat, bought especially for the occasion.

The train announced its arrival with a hiss of opened valves and drew to a stop amid a cloak of billowing steam. A tall figure appeared. Käte made a small sound and leapt into motion, black hat flying from her head.

Her husband opened his arms and she slammed into him.

'I thought I'd never see you again.' She buried her head into his chest.

Freundlich encircled her with his arms and rested his head on hers.

Einstein scooped up her hat and turned away, their reunion evoking strange feelings. He found himself thinking of the letters he had written to Mileva years ago, after his graduation had forced them apart.

She had been back in Serbia with her parents and he had been in Italy with his. Mama had been doing her best to dissuade him from pursuing Mileva, convinced that the Serb was beneath him racially and intellectually. He had poured out the injustice of his predicament to Mileva in letter after letter of ungloved passion, until finally they had been briefly reunited in the spring sunshine of the following year at Lake Como.

He never did see the daughter that their reunion had produced. Lieserl breathed her last following a bout of scarlet fever shortly after birth and her body now lay in Mileva's hometown, in a grave he had never visited.

'Oh, you are kind,' said Käte, interrupting his thoughts as she retrieved her hat from his hand.

Freundlich was watching him curiously. 'I must apologise that I haven't brought back any results.'

Einstein shrugged. 'Don't trouble yourself. The effect is in no doubt in my mind, theoretically at least. The observations are just detail.'

'But surely you need confirmation?'

'Only to convince others,' said Einstein with bravado.

There was a young blond man behind Freundlich, gawping around the station as though he had never seen it before.

'I can't believe I'm home,' said Mechau, grinning sheepishly. He turned to shake Freundlich's hand warmly and, with a few final words of friendship, disappeared into the crowd.

Freundlich reached down and took Käte's hand, saying, 'Time for us to leave, too. Thank you for coming today, Albert. You may remember that I paid you a lot of attention on my honeymoon, so you will forgive me if I do not make the same mistake today.'

Einstein forced a smile. 'Of course. You know where to find me.'

The couple headed off.

'Let's talk soon,' Einstein called after them.

★

Walking briskly to keep out the cold, he entered Alexanderplatz and caught the sound of a woman's cries over the noises of the trams and the people. She was rail-thin and had collapsed beneath the casualty lists posted up on the side of a building. An ashen-faced man was trying to comfort her and lift her back to her feet, but her spindly limbs had no strength in them. The more he tried, the harsher became her cries of anguish.

Another man, elderly in old-fashioned clothing, was berating the woman for treacherous behaviour, telling her that her son had died for the greater good of the Fatherland. 'Why can't you be content with that? You should be rejoicing!'

The husband managed to haul his wife to her feet. He gave the other man a beseeching look but the tirade continued.

'Look at me. Both my sons have died in Belgium and I'd gladly have died with them. It would have be an honour.'

The woman looked suddenly murderous. 'I would gladly exchange my life for his. What do I have to live for now?'

Einstein saw the sting of it on her husband's face. What happened when a wife turned into a mother? It was bad enough when they turned from lover to wife, but upon the first child the husband became almost irrelevant.

Einstein's own assessment shocked him. He did not hate women. Did he? No, it was Mileva poisoning his mind. Today's letter had been hectoring: had he remembered that his next payment to them was due? – *Yes!* Would he write to Hans Albert? – *I do!*

He turned away and hurried across the square. He needed to see Elsa and prove to himself that he was not as cold-hearted as he suddenly feared.

She looked overjoyed to see him, but danced out of his way and proffered only her cheek when he stepped close to embrace her. 'This is a surprise, Albertle. I have somebody here I'd like you to meet.'

'Meet?'

'Yes, meet. He's in the sitting room.'

Einstein felt his plans slipping away. The affectionate chat, the compliments he had planned to offer as balm against his decision to stay in his own apartment. Perhaps he could have coaxed a giggle from her. That would have been a coup. She usually met his humour with a shake of her head, even though he could see from her shining eyes that she wanted to laugh.

He entered the parlour and a man yelped in surprise and jumped to his feet. The reaction was so sharp that Einstein wondered for a moment what he had interrupted. The startled individual was tall and slim, dressed in a well-tailored three-piece suit of charcoal-coloured wool. Einstein estimated that they were of similar ages, yet the stranger was rendered younger by not having acquired much in the way of self-confidence. He wiped a hand along the sharp line of his dark trousers and thrust it forwards. It was clammy in Einstein's grip.

'Praise God you've come, Herr Einstein!'

'I hope that's metaphorically speaking,' said Einstein warily.

The man floundered immediately, looking around and blinking furiously.

'Now, now, don't tease him with your own lack of faith, Albertle. Take no notice, Georg, he was only teasing,' said Elsa with a warning look.

Actually, I wasn't, thought Einstein.

'This is Georg Nicolai, he's a friend of Ilse's and a doctor at the university,' she continued.

'Physiologist, really,' he said apologetically to Einstein. 'Sir, I came here today to ask if you have seen the paper this morning.'

Einstein shook his head; this was becoming more tedious by the moment.

'Here.' Georg handed him a newspaper, folded open at an article entitled *Appeal to the Cultured World*.

As representatives of German science and art, we hereby protest to the civilised world against the lies and calamities with which our enemies are endeavouring to stain the honour of Germany in her hard struggle for existence – in a struggle that has been forced on her.

Einstein read in disbelief. It was nothing but a list of excuses and double-talk for the atrocities being perpetrated in Belgium. It was common gossip that the wanton destruction at Louvain was a means of teaching the Belgians to respect Germany. Yet here the academics were claiming that only small parts of the city had been set alight and that the deed had been undertaken by troops with 'aching hearts', to punish the civilian population for their resistance to occupation.

'Who are these representatives of German science? I was not approached!' Einstein exploded, making Nicolai flinch.

There was a long list of signatories: Haber and Nernst he could have predicted, but one name shocked him. *Planck!* He flung the paper to the floor. 'Outrageous!'

Nicolai edged forward. 'May I show you my response?'

'How can you respond to that?' Einstein shot a finger at the scattered sheets on the rug.

'With a counter-manifesto, signed by other representatives of German science and art. We will call for educated men of all countries to come together and help to end this war. There must be a united Europe in which even the possibility of war does not exist. There can be no victors in this war. Either Germany will gain new ground from countries that will forever resent our presence, or we'll be so heavily defeated that our people will suffer for years. Either way, it breeds more contempt and more reason for future wars. We must find a way to stop it once and for all.'

Einstein looked into the man's face. 'I misjudged you, Georg. I'm pleased to have made your acquaintance.'

Georg's eyes darted away in embarrassment. He bashfully fumbled with a sheet of paper, handing it over. Its message was written in elegant handwriting. 'Please edit in any way you see fit. We are not the Europe of four hundred years ago, ready to go war on a whim. Battles are no longer fought on fields but over entire countries. Civilians are being slaughtered by fully armed troops . . .'

'Save your words, Georg. Don't look hurt, I mean you no disrespect. I want you to save them for those who need to hear them. I need no convincing. I'll take your document to Max tomorrow for signing.'

'You'll take it to Planck?'

Einstein nodded.

Nicolai's doubtful expression accentuated his sharp features. 'But he signed the manifesto.'

'No! Max would never knowingly put his name to such a document. You don't know him as I do. He must have been duped.' Einstein brandished the manuscript. 'I will take your work to Max myself and get him to sign. You'll see what he really thinks then.'

Planck looked as if he had not slept in a week. His face was haggard and his eyes were darkly ringed. He kept lifting his glasses and squinting at the document Einstein had thrust in front of him. As he read, he seemed to shrivel into the office chair. Einstein rocked expectantly in front of the desk, hands clasped behind his back.

At length Planck laid the sheet of paper on his desk and smoothed it with his hands.

'I cannot sign this,' he said sadly.

'Why ever not?'

'I cannot put my name to something I do not wholeheartedly support.'

'But, Max, you signed that vile piece of work in the newspaper.'

Planck's chest heaved; he unwound his wire glasses from his ears and massaged the lobes. 'Yes, I did.' His voice was scarcely more than a whisper.

'Then why not this one?'

'Because I believed in what it said.'

Einstein shook his head. 'Impossible.'

Planck looked at him. 'You don't know the full story of Belgium, Albert. You're too quick to assume German villainy, but Belgium had agreed to the occupation of its lands by British and French forces to defend it against Germany. We would have been mad not to pre-empt that threat.'

'How do you know that?'

'If you listened to the discussions at the Academy instead of storming off, you would know it, too. Some of the members are very well connected. The French are using dum-dum bullets to rip our boys apart despite it being specifically forbidden by international law.'

'And that's why it has to stop. I saw a woman break down in the street a few days ago. Her son had been killed and she was inconsolable, couldn't even accept the comfort of her own husband. Last night I was writing to Hans Albert and I realised what it must be like to lose a son. Not just to have one taken to another country, but to lose him for all eternity.' Planck was absent-mindedly stroking his drooping moustache. Einstein plunged on. 'We're both family men. Our sons – everyone's sons – are a precious commodity. They shouldn't be wasted in slaughter . . .'

'Erwin is missing,' Planck said quietly.

'No, he's back. I've seen him myself. He . . .'

Planck shook his head slowly. 'Not Freundlich, my Erwin. They think he was taken by the French.'

'Then, Max, how can you defend the war?'

'We are cast as demons by the old powers. England, France, Russia; they don't want us challenging their empires. So, they write fictions about our motives and us.' Planck's voice grew in volume. 'A strong Germany is a threat to them; they'll stop at nothing to keep us down. They're the aggressors, not us. We're striking back against them before they have the chance to attack. I hate the war as much as you. But it is a just and honest war. I'm not going to insult you by saying it's glorious. We both know war can never be that; but sometimes it can be inevitable and necessary.'

'Even if it comes at the price of your own son?'

Planck was stony-faced. 'Even if it comes at such a price,' he said through clenched teeth.

'These past weeks I have clung to the notion that all Europe has been going mad around me. But now I sense there has been madness in me as well. I thought I could just sit it out and wait for everyone else to come to their senses. How could I have been so naive? Surely, the madness is for me to let it happen unchallenged?'

Planck replaced his glasses and held up Nicolai's manifesto. 'I will not sign this. Now is not the time for apologies. Don't alienate yourself, Albert. Not now. It'll only reflect badly on you after the victory.'

The unexpected steeliness in Planck's voice ignited Einstein. 'Was that a threat?'

Planck's round eyes took on the expression of a disappointed father. 'You have to stop pretending that you're bulletproof, otherwise there'll be no place for you in Berlin.'

Einstein snatched the sheet from Planck's hands. 'Perhaps it was a mistake to come here in the first place. I was under the illusion that you were one of the most intelligent men in the world.'

The wind was bitter that evening when Einstein stopped at the flower stall near the station. This late in the year the selection was seriously depleted, and prices were high from having to transport the stock up from Italy. He selected a single red bloom, handed over some money to the shivering flower-boy and hurried on his way, catching snatches of conversations from the other pedestrians.

'Of course it will,' one man was saying. 'Victory in six months, mark my words. There'll be a big push, everyone's talking about it.'

Einstein blew out an angry breath and crossed the road. The street lamps glowed feebly, drooping like huge dewdrops from their iron armatures.

A German victory would be a disaster. It would embolden the Kaiser's imperial ambitions beyond imagination. Einstein could just hear the crowing and the arrogance. It was unthinkable.

But the alternative . . .

One only had to read between the lines to see the truth. The longer Belgium resisted, the more entrenched the front-line would become; there would be stalemate and standoff. Germany would be isolated and the country besieged.

Einstein did not know which conclusion to hope for. Victory or defeat: each was as appalling as the other.

He was still mulling over the awful possibilities when he arrived at his destination, a tall house with curtains so heavy against the coming winter that there was no sign of life inside. He straightened his shoulders and worked the bell-pull.

Max Planck opened the front door. He bristled at first sight of his visitor, but then his posture softened as he saw the red flower and his face filled with understanding.

Two summers before, Haber had dispatched Planck and Nernst to Zurich to woo Einstein to Berlin. He could still remember the dizzy feeling as they laid out their offer in the busy station café: a professorship at the university with no teaching duties but the highest salary possible; a membership of the Prussian Academy of Sciences, carrying a stipend; and the promise of a directorship of his own institute of physics once funds became available.

They told him how Berlin was growing so rapidly that it would soon be the engine not just of Europe but of the world. All he had to do to share in this new power was leave Zurich and its mountain walks behind.

But Einstein had understood the personal price he would end up paying; he had seen it that morning as clearly as he had watched the waiter mop the bar. Living in the same city as Elsa would be too much. His marriage to Mileva was already strained; having his cousin around would lead to its disintegration. The letters he was already exchanging with Elsa convinced him of that.

So, he asked for time, conjuring looks of incomprehension on the faces of his tempters.

The next morning he had still not made his decision. In the moonlight he had studied Mileva's face and seen enough in her soft contours to remind him of why they had originally fallen in love. Yet the lure of Berlin was an almost irresistible force.

To give himself more time he had packed off Planck and Nernst on a train journey that took them winding through the scenic foothills. On their return, he told them, he would be carrying a single flower. No words would be necessary. If it were white, he was staying with the snowy peaks of Zurich. But if it were red, he was changing his life irrevocably. He would not just move to Berlin, he would declare his intellectual allegiance to Planck, Nernst and Haber.

Back that summer, with the sun so high that he could see Nernst hanging out of the carriage window before the train was anywhere near the platform, he had greeted them holding the red flower of acceptance.

Nernst had raised his arms in triumph. Planck's eyes had creased with a smile and a single nod.

Now it was winter, eighteen months later, and everything was different. Nevertheless, Einstein held out another red flower.

Silhouetted in the yellow lamplight from his hall, Planck looked at the proffered bloom. 'Come in, my friend,' he said, 'you must be frozen.'

7

Diksmuide

Physics textbooks were not the only reading matter Lemaître had brought with him from Brussels. He also had his Bible. When he was sent back from the front line to snatch some rest, his thoughts were normally too frenetic to allow him to sleep right away. Only if a battle had worn on for the whole day would he be exhausted enough to sink straight into unconsciousness, but then his dreams would fill with the day's horrors. So, whenever he could, he read silently by the tent in the flickering firelight.

Once some sense of equilibrium had returned, he would pray for the souls of his fallen friends and for his brother's safe deliverance. Jacques had been stationed further north, up along the Yser River, where the gossip said the fighting was just as fierce. Lemaître would hear tales of shells falling into the river and producing waterspouts dozens of feet high that soaked the troops, leaving them to dry out in the bitter winds. Jacques hated to be wet, and used to complain all the way home if the clouds had burst while they had been out cycling.

Then would come the most difficult part of Lemaître's nightly ritual. He would replay the enemy casualties he had witnessed that day and pray for their deliverance too.

Sometimes he saw them fall to the ground, animate one moment but leaden the next. At other times, he noticed them just as bundles of clothing in the mud, or sprawled at awkward angles. Occasionally, they would simply vanish in the explosions. It was impossible for him to imagine the shock to their souls as they were ripped from their mortal lodgings, and so he relived the moments of their deaths and asked God to welcome them.

Otherwise, he reasoned, there would be no one praying for them for days, perhaps even weeks, because of the time it would take for the news of their demise to filter home. There would be parents going about their days in ignorance because the lines of communication were so slow. When the moment came for them to open the telegram and for the pain to cut into them, their son would have been dead for weeks.

Only God could know the exact state of the world as it happened, thought Lemaître. Human perception would always be thwarted by physical limitations. So he prayed on behalf of all the anonymous parents, while all the time thinking about his own.

German troops now attacked daily, bolstered by reinforcements, whereas Lemaître and his colleagues were being whittled away with each bullet and bomb that fell in their midst. Soon they would be a skeleton force. Then one morning, new noises joined the battle.

There was a heavy bark of guns, deeper and more menacing than anything Lemaître had heard before. He looked along the dyke and saw the unspoken question on other faces. Only the sergeant was grinning, preening his grey moustache.

'It's the British: they've got warships anchored off the coast. Those rounds'll plough up the German lines for us and no mistake.' He sealed the good news with a wink.

Inevitably the Germans answered in kind, but their weapons could not possibly reach the ships, not even the giant howitzers they had rolled up a few days earlier. So, as had become the norm, they concentrated on the defensive lines in front of the town. Yet as the day wore on, there was no doubt in Lemaître's mind that the bombardment had lessened. Even the grenade attacks were more modest, designed it seemed to harass rather than break through.

Occasionally the enemy gunners would lift their sights and target the town itself. Lemaître would watch from the temporary safety of the frozen earth as one building after another was chipped away. When a shell really found its mark a whole building would collapse, sending billowing clouds of brick dust through the streets and thick fingers of smoke up into the sky.

Yet it all felt rather desultory, and Lemaître found himself warming to the notion that perhaps the invaders had lost heart. Those with field glasses had a different story to tell. The bombardment had shifted merely northwards, intensifying the battle along the river. Lemaître felt helpless and spent the afternoon fretting about Jacques.

The night battalion arrived on schedule as the town was turning into silhouettes. Now that the battle had moved away, the raucous accompaniment of the guns had been replaced by an unsettling quiet. The only sounds were coming from the battalion dogs, roused to barking by some unknown phantom.

'All right, boys, let's get ourselves out of here for the night.' The sergeant counted them out of the trench with a slap on each shoulder and the squad withdrew under the light of a heavy rising moon.

Lemaître and the others trooped off, keenly aware that their backs were turned to no-man's-land. They knew better than to relax before they were well hidden by the buildings towards the rear of the town.

Lemaître was about to round the corner into the market square when noise assaulted his senses. Sudden and absolute, it was as harsh as a physical blow. He hit the ground, unsure whether he had been blown off his feet or dived instinctively, and grabbed his helmeted head to hold the protective head-gear in place. He buried his face in the dirty street, eyes squeezed shut, and felt the rain of debris pummel his body.

Dust filled his nostrils, threatening suffocation. His eyes sprang open and he hauled himself to his feet. Piles of rubble and bricks surrounded him. Moonbeams cut through the skeleton of what had been a home until a few moments ago, their silver light slipping through a shattered ribcage of broken banisters.

His knuckles were smarting, and a quick inspection told him they were badly skinned. He flexed his fingers and then wafted his hands in the cooling air to soothe them.

Others were rousing too, spectres moving through the settling clouds of pulverised mortar. Lemaître mentally counted off the faces one by one.

'Where's the sergeant?' he called.

They organised a search and began combing the debris. Lemaître used his feet to topple any pile that looked big enough to hide a man.

'He's here,' came the shout.

Lemaître slithered over the uneven debris to where they were hauling out the sergeant by his armpits to prop him against a wall.

Lemaître squatted to study the avuncular face, hoping for some glimmer. There was nothing except a frozen look of profound annoyance. Lemaître bowed his head and prayed. Whatever had made the sergeant more than just flesh had already departed.

After a fitful night, Lemaître and the remains of his squad were ordered out of bed earlier than any of them had feared. There was a sense of urgency bordering on panic in the town. Men with shovels were reinforcing the defensive mounds and soldiers were forlornly counting their ammunition, looking around to see if they could possibly have dropped any.

The news was not difficult to discover. Yesterday's offensive had been successful; the Germans had broken through further north and crossed the Yser. Diksmuide could be surrounded in days.

'We're done for,' one of the squad muttered, and spat on the floor.

Lemaître and the rest were hurried to a schoolroom where the small desks had been shoved into a pile so that the men could crowd in.

A map was pinned over the children's drawings that lined the wall. It showed the bridgehead of Diksmuide, sticking out like a nose to the east of the Yser. The waterway ran more or less north to Nieuwpoort, where it emptied into the North Sea. To the west of the river was a railway, marked by a line that curved away from Diksmuide on a circuitous journey to the port. A blunt pencil had been used to hatch the crescent-shaped lowland between river and track. The commander used his filthy fingernail to tap this shaded area.

'This is the last line of defence. Flood the polders here, and we'll save Nieuwpoort.'

Flood the polders! Lemaître could hardly believe his own ears.

The commander pressed on, speaking quickly. 'The Germans will have no choice but to turn southwest, and that will take them straight into the greater assembly of allied forces. There's a full moon in two days' time that will bring a high tide capable of turning these fields into a swamp. We must be ready for the coastal sluices to open. And that means that we have the most important duty of our lives to perform.'

He swept his gaze around the grubby soldiers, making brief eye contact with every man in the room before continuing.

'The railway is built on an embankment containing twenty-two culverts. These keep the fields drained of excess water and must therefore be blocked before the sluices are opened. They must be completely blocked if the flooding is to succeed. No mistakes.'

Lemaître's body prickled. He thought of the great machine of the German army pressing down on them. It was relentless, virtually invincible; supplied directly from the Rhineland, rotating its troops away to rest for days or even weeks at a time; it was impossible to beat man for man, but damming the culverts so that the sluices could be opened – that could be done.

'Lemaître!'

He stepped forward automatically. 'Yes, sir.'

'I hear you were studying engineering before you volunteered. You'll be in charge of a squad.'

'Yes, sir.' Now was not the time to confess that he was veering towards physics because the mathematics needed for engineering had proved too simple to hold his attention.

At dusk, the squad of a dozen men made their way along the tall railway embankment. Rifles were strapped across their backs, shovels and pickaxes stowed in the dogcarts rattling along beside them.

The fields were bare after the autumn crop and hard underfoot from the frosts. If the plan succeeded tonight, it could be years before these lowlands were drained again.

Lemaître called his squad to a halt at the first culvert and peered inside. The drainage channel was only tall enough for a man to crouch low, but it was deep, extending about five metres through the embankment beneath the track. They would not have to fill its entire depth but they would have to completely block the entrance. He estimated it was probably two hours' work, and thought briefly of the other squads sent to other sections of the line.

The men swung their picks but they made little impression in the frozen soil. Lemaître raised his estimate for each culvert to three hours; it was going to be a close thing. Around him the icy air carried the sounds of distant battle. He moved his shovel quickly, urging his men to do the same, and the culvert filled steadily.

As the night fell, a false dawn of battle-fires and the white flash of explosives lit the eastern horizon. He glanced around and saw another light, much closer. It was the yellow glow from the window of a small, whitewashed farmhouse sitting in the middle of the field. The moonlight caught hold of a thin grey twist of smoke rising from its chimney.

He sighed; he would have to warn them to leave. As he passed the dogcarts, he noticed the dogs quivering with cold.

'Exercise them,' he ordered a recruit. 'We can't risk losing the dogs.'

'Yes, sir.'

Lemaître cut off over the field until he arrived at a worn track that led to the clapboarded farmhouse.

He knocked on the door, calling, 'I'm a friend, a Belgian.'

The door opened to reveal the farmer holding a cumbersome old shotgun, levelled and ready. Lemaître jumped back, hands raised. 'Please! I've come to warn you. This whole area's to be flooded.'

The farmer's eyes narrowed.

'It's our only hope. The enemy'll be here tomorrow or the day after. We have to flood the polders to stop their advance or all of Belgium will be in their hands.'

A frightened woman, huddled in a shawl, appeared from behind her husband. 'But this is all we have.' Her desperation was palpable.

Lemaître turned for the gate. 'I'm so sorry, but you must go. I will pray for you.'

When he returned to his men he found that they had made good progress; some had stripped off their woollen jackets to work in their shirts or vests.

'Good work, lads. I'll have to leave you to your own devices more often.'

Lemaître helped as they delivered the last spadefuls and set about compacting the earth with the flats of their shovels, fashioning an earthen buttress over the drain. Lemaître checked his watch: two hours and fifteen minutes. He nodded in satisfaction: better than he had hoped.

'Let's move on.'

They were working on the third culvert when a rickety cart, pulled by an old nag, creaked past. It was piled haphazardly with belongings, including an upturned rocking-chair and a bundled clothes-line. Lemaitre recognised the occupants and watched them pass, hoping for some acknowledgement, but the farmer and his wife remained stony-faced, their gazes fixed firmly ahead.

By the time the squad reached the final culvert, it had become increasingly difficult to walk. The ground was clawing at them and caking their boots in mud.

They've opened the sea gates already, thought Lemaître. 'No time to slow down now, lads,' he called. 'We're nearly done. The battle's nearly over.'

They set to with a renewed urgency and, having completed the task, marched the final distance into Nieuwpoort. The nearer they approached, the more swamped the conditions became. Before long the dampness began to climb the wool of Lemaître's uniform trousers. Around him ovals of water dotted the furrows, shining silver in the moonlight, and the dogs whined as the carts became mired.

He had no choice but to order the men to lift the carts and carry them on their shoulders. Others took charge of the dogs, allowing them enough leash to bound over the worst of the quagmire.

When they finally trudged into Nieuwpoort, Lemaître thought he would sleep for a week. He reported in at the sentry post and arranged for the dogs to be kennelled. Then he followed orders and ushered his squad into a church hall so that they could be debriefed.

They had a day's rest before the battalion commander again summoned them to the hall. He looked exhausted, his tiny eyes half shut, and Lemaître feared the worst. Yet when the commander spoke, he said, 'Congratulations. It worked. The Germans have broken off the attack.'

Around him the cheering reached to the rafters. It was as if they had won the war. The men slapped each other on the back, punched each other on the arm and mussed one another's hair.

Lemaître raised his voice too, but then his mind filled with images. Sinking into the mud, the Germans would have been sitting ducks. It did not take long to find those in the camp who had seen the ensuing slaughter. Regrouped and ready, the Belgian troops had killed thousands of men until the Germans had sounded the retreat. Lemaître would be busy with his prayers tonight.

8
Berlin

The night was turning to stone around them as they waited on the top step in front of the Habers' front door. Even through his woollen overcoat, Einstein could feel the icy barbs touching his skin. He stepped forward and rang the bell a second time. As before, no one answered. He turned to Elsa with a shrug.

'Are you sure you got the right night?' she asked.

He ignored the question. He was already on edge from having agreed to take her with him. Their relationship was the worst-kept secret in Berlin, yet he still felt nervous about publicly acknowledging it.

She chattered on. 'You know what you're like for forgetting things.'

He pointedly ignored that comment too. 'No lights on.' He gestured towards the house. 'What could have happened?'

'Well, we can't stay here. It's freezing.' Her fleshy face was already pale with cold.

'You're right. Let's leave a note and be on our way.' He began patting around his pockets for a scrap of paper and a pen when a squat figure appeared from the night, waddling hurriedly along the street. 'Walther?'

Nernst stepped fully into the lamplight.

'Walther, when did you get back? I thought you were still at the front.'

Nernst's breath came in giant gasps. 'There's been an explosion at the university – in Fritz's lab. Someone's been killed.'

'Killed?' Elsa shrieked.

Einstein scuttled down the steps. 'Where's Clara?'

'She's gone over there. To the university.'

'Then we must go too,' declared Einstein.

'Are you sure that's wise?' said Elsa uncertainly.

Einstein was already hastening off down the street. 'Go home and wait for me. Walther, what was Haber working on?'

An uncomfortable expression crossed Nernst's face. 'Perhaps best you don't know.'

There was a cordon blocking their way fifty yards from the laboratory. Einstein hesitated until he glimpsed Clara, down near the entrance. He ducked the rope and approached. Her arms were clasped around her body. Her lips were drawn into a tight line and she was staring at the wooden blocks of the corridor's floor. Einstein feared the worst and quickened his pace. Nernst jogged along beside him.

Einstein almost shouted in relief when he caught sight of Haber. He was standing in the doorway to the chemistry lab, contemplating the wreckage. A desk had been turned to matchwood and there was extensive charring on the ceiling. The blast had smashed all the windows and scattered a number of metal canisters across the floor.

Haber barely acknowledged them. He seemed haunted by the presence of his wife, who flicked Einstein a look but said nothing. She was studying her husband so closely that it was clearly not with concern but something more hostile.

Haber stood aside as a pair of men in medical uniforms carried out a stretcher. The person on it was anonymous, covered head to foot by a white sheet.

'Poor Detlev,' moaned Clara.

Haber swung to glare at her. 'He was killed outright. He wouldn't have known a thing.'

She met his gaze with bright contemptuous eyes and he turned his back. He began to clear away the canisters, handling them as carefully as newborns.

At the sight, Einstein felt a chill, colder than anything the night had to offer.

Einstein was already sitting in the restaurant, studying a simple typed menu, when Freundlich burst through the entrance, setting the door blind rattling against the glass. The two men exchanged greetings and soon the astronomer was reliving his captivity as if it had been some kind of Boy Scout misadventure. 'I don't mind telling you, when they called my name from the ranks, I thought I was a dead man.' He told how his group had been led away by soldiers but, instead of facing a firing squad, they had been shipped off north-eastwards – he could tell that from the stars through the carriage windows on the guarded train. Only when they were taken to the German embassy had he realised that they were being repatriated.

'The observatory is virtually empty these days,' said Freundlich. 'Even Karl has volunteered for the army.'

'Schwarzschild? I don't know him that well.'

'You should get to know him better; he's an expert mathematician. I think he's very interested in your work.'

Einstein laid down the menu. 'Then he's the only one who is. Most of them dismiss it.'

'Is it going badly? I thought you were so close.'

'So did I.'

The blind alley he found himself in had become obvious just a few days before, when he had realised that the equations he was using to describe the curvature of space would not yield the correct precession of Mercury's orbit.

'I need to start again with better equations,' said Einstein, 'find a set that makes the answer the same, regardless of the observer's state of motion. Acceleration is the key to this. But for the moment, I'm back to square one.'

The waiter appeared and Freundlich hurriedly picked up the menu. Pork was the most obvious and cheapest thing. The authorities had started slaughtering pigs because their feed took food away from the human population. The slaughter was also serving to put cheap meat on the table again, or on non-Jewish ones at least.

'Do you?' asked Freundlich.

'I've eaten pork since I was twelve and my intellect began to question blind assumptions,' replied Einstein, and ordered the pork belly.

With a look, so did Freundlich. It came in heavy gravy with a rather meagre lump of bread.

'At least a quiet observatory must be a good place for concentration,' said Einstein.

'On the contrary, my duties are doubled. I'm doing all their work as well. I have no time to think about anything any more.' There was clear resentment in Freundlich's voice. 'Albert, if something doesn't happen – if you do not help me – I fear that I may never find the time to organise another eclipse trip.'

'When is the next eclipse?'

'The third of February, 1916.'

Einstein grimaced. 'Too early. This hateful war will never be over by then. It'll soon be Christmas as it is. When's the next one?'

Freundlich recited them from memory: the eighth of June, 1918, and the twenty-ninth of May, 1919.

'Not until 1918?' Einstein's thoughts were turning like cogs.

'But we must at least start getting the equipment back, all the cameras.'

'I'll try, but don't get your hopes up.' Einstein muttered.

Freundlich leaned across his meal. 'I thought you needed the eclipse data to finish the theory?'

'Not really. I can use the Mercury measurements to fashion the theory, from which I can then predict the deflection of the starlight. Imagine if I could predict the deflection. No one would be able to refute my ideas then.'

Freundlich's face lit up. 'It would be a perfect Newtonian experiment – a crucial experiment. As important as when Halley used Newton's theory of gravity to predict the return of his comet.'

'Exactly!'

'You make the prediction, and I'll take the measurements.'

Einstein wanted to laugh. 'It's a deal.'

The strained atmosphere between Haber and his wife had not eased by the time Einstein and Elsa arrived at the house the following week for the rescheduled dinner party. Haber and Clara sat at opposite ends of a polished redwood table, doing their best to engage their guests in conversation, and Einstein found himself wondering why they had gone ahead at all.

Nernst was there with his wife, Emma. Each was as plump as each other and apple-faced. They insisted on sitting next to one another and spent the evening sharing asides and the occasional playful touch. The only other guest was a friend of Haber's. Walther Rathenau sat between Einstein and Elsa, opposite the Nernsts. He was so squarely in the middle of the gathering, it would have been easy to mistake him for the host.

He swaggered even when he was seated. It was in the way he reached for the salt or lifted his crystal goblet to taste the wine. His eyes took in everything, unblinking, set below heavy brows and his close-cropped salt-and-pepper hair.

Although Einstein did not know him, he recognised his name from the papers. He was an industrialist who had been appointed to the War Department.

'I'm to keep Germany in this war for as long as it takes,' Rathenau explained.

Haber lifted his glass. 'To victory!'

The others, excluding Einstein, followed suit. 'Victory!'

Elsa looked mildly embarrassed that she had joined the toast. She fiddled with her napkin before turning to Clara for distraction. 'Last week's accident must have been awful for you. Did you know the young student well? Have you had the funeral yet?'

Einstein could not believe his ears. On the walk over he had expressly told her not to raise this subject.

'Hardly at all,' admitted Clara, but that did not stop her throwing a filthy look across the silver tureens at her husband. 'And he was buried yesterday.'

Elsa was just opening her mouth to ask another question when Clara asked Einstein: 'Have you heard from your wife lately?'

Elsa winced.

'I have heard from *Mileva*, yes.' Einstein placed a heavy emphasis on the name, hoping to ward off further uses of the word 'wife'. 'It worries me that she has not yet secured a fixed address. She left Berlin four months ago now.'

'She still believes that you'll come to your senses and reunite.'

'Then she is labouring under a delusion. I am content with my own company.'

Immediately the room's attention shifted towards Elsa. She had a hand to her face, not knowing what to do. Einstein's cheeks burned and Clara skewered him with a look. 'Will you be seeing your children over Christmas, Albert?'

He bristled. 'I will be spending Christmas alone.'

His statement brought a look of naked incomprehension to his cousin's face. He gabbled on, trying to talk his way out of the situation. 'Elsa has a full house, what with her two daughters and her parents visiting. It would not be fair to impose myself.'

'You can't spend Christmas alone,' said Nernst.

Einstein forced himself to laugh. 'Listen to us, all of us Jewish and discussing how to celebrate the birth of Christ.'

Nernst and his wife conferred for a moment. 'It is decided. Albert, you are coming to us.'

Utterly trapped, Einstein quietly accepted. Out of the corner of his eye, he could see that Elsa was close to tears.

'Albert, that was a good presentation you gave to the Academy the other week,' said Haber hastily, attempting to force a more casual conversation.

'Did you think so? I can't shake the feeling that I'm nothing but an irrelevance at the Academy these days.'

'Perhaps it would help the Academy if you explained your original paper on relative motion more fully,' said Nernst, 'instead of just blundering on with these new ideas about gravity.'

'Blundering on? You weren't even at the meeting.'

'You have to admit relativity is very difficult to understand.' Nernst had a smarmy grin on his face.

'There's nothing difficult about relativity,' said Einstein impatiently. 'That first paper was all about how we view events relative to one another. Common sense tells us that we should all see the same event at the same time, but that's not strictly true.'

'We are simple chemists, be gentle with us,' pleaded Nernst, raising a giggle from his wife.

Einstein pretended not to have heard. 'Imagine a moving train with two lights, one on the front and the other on the back, synchronised to flash at the same time. Now, imagine two people watching those lights: one standing inside the carriage, halfway along, and the other standing on the embankment. The question is: do they both see the lights flashing simultaneously?'

His answer was the scrape of cutlery against plates. He was about to continue when Rathenau lifted napkin to mouth and dabbed. 'I'll answer if no one else is bold enough. Yes, they both see the lights flashing in time. How can they not if the event is simultaneous?'

Einstein nodded vigorously. 'Exactly what Isaac Newton would have said.' He stopped his nodding. 'And completely wrong.'

Rathenau laid down the napkin.

'The person on the embankment does indeed see the flashes simultaneously, because the two paths taken by the light beams from the forward and rear lights are essentially the same distance,' explained Einstein. 'But the person in the carriage sees the lights flashing out of synch because, in the time it takes the beams of light to reach him, the train has moved. It's carried him forward a little, meaning that the light from the front of the carriage has less distance to travel to reach him. The opposite is true for the rear light, which now has a little extra distance to travel. So, the forward light reaches the observer on the train a little earlier than the rear-ward light. Hence, the two observers cannot agree on the timing of the light pulses unless . . .'

'That's madness,' said Rathenau.

'No,' said Einstein with a broad grin, 'that's physics.'

The industrialist pulled his goatee beard into a point. 'So, if the Czar of Russia were on a train and an assassin throws two sticks of dynamite, one to the front and the other to the rear, you're saying that what is but a single moment of danger for the assassin ends up startling the Czar twice.'

Einstein was taken aback by the darkness of the example but nodded. 'They will never be able to agree on what they see unless they take their difference of motion into account. The man on the embankment is stationary; the man on the train is moving. My first paper shows how to do this.'

'And I gather from tonight's conversation that you're trying to extend this?'

'Yes. My first paper dealt with stationary observers and those who are moving with a constant speed, but it doesn't work for observers who are changing speed. Now I want to find the general theory, the one that works for accelerating and decelerating motion as well, and when I have discovered that, it will open a whole new way of thinking about gravity.'

'Gravity?' There were the beginnings of a pained expression on Rathenau's face. 'As in Isaac Newton?'

'To Newton gravity was a force created between all objects with mass that pulled them together or spun them into orbits, but he didn't know what gravity was. In fact, at one point in his life he suggested that it might even be the will of God.'

Rathenau snorted.

'I'm working to explain gravity not as a mystical force conjured through space, but as a property of space itself; as a curvature – or a contour, a warping, a valley pressed into the fabric of the universe. Each celestial object makes its own impression on space, and then smaller celestial objects move according to these contours. My general theory will be able to give the shape of these contours depending upon the size and mass of the objects.'

'Can we see the curvature?'

'No, but we feel its effects all the time through the acceleration of objects pulled by gravity. By understanding accelerated motion, I can understand gravity in a way that Newton couldn't. I can explain things that Newton's theory can't, such as the orbit of Mercury.'

'You're planning to dethrone Isaac Newton.' The flat tone of Rathenau's voice made it hard to decide whether the statement was intended as admiration or derision.

Einstein raised his chin. 'I thought challenging British authority was what this new age was all about.'

The table fell silent. Rathenau's eyes bored into him. 'I like you, and your ambition, Herr Einstein. There's just one thing I don't understand . . .'

'Only one?' joked Nernst, looking up from his empty plate. 'You should be careful. If you let him, soon he'll be telling you about how an object travelling at close to the speed of light will appear shorter to an outside observer.' He rolled his eyes to the ceiling.

'So, you have been listening,' began Einstein.

'Oh but Walther, it's easy,' cut in Haber. 'He's saying that if you run fast enough, you'll appear thinner. Surely that's a good thing for you.'

Nernst clapped his hands to his belly. 'If I could run fast enough, I wouldn't have this in the first place.'

Einstein slightly raised his voice. 'That's not the only consequence of such rapid motion, and the general theory takes things further. It predicts that the gravity of the sun will deflect passing beams of light . . .'

'Albert, please! You're not in the Academy now.' Haber framed the remark as a joke, but his tone contained an undercurrent. 'Think about the ladies you're boring.'

Clara made a scornful sound. 'We're not bored listening to new science. How could you think that?'

'Well, I am,' said Emma. 'It's over my head.'

'There. Thank you,' said Haber.

'Why not tell everyone about *your* work, husband?' Clara said. Haber glowered at her. 'I'm sure we all want to hear your plans for shortening the war.'

Einstein pictured Haber and his canisters in the lab.

'Someone has to bring this stalemate to an end,' their host growled.

'With poisonous gas?' Clara's voice was glacial.

There was an awful silence. Every eye was on Haber.

'It's quick, it's humane, and it will save lives in the long run by shortening the war. Even Albert here has done war work. Oh, don't look at me like that. I know about that compass design you patented and sold to the shipbuilders in Kiel.'

Einstein glared. 'That was nothing like what you're doing.'

'It'll be effective in tracking down enemy ships.' Haber looked to the other side of the table. 'And you, Walther, you've worked on gas weapons.'

Nernst squared himself indignantly. 'Non-lethal gas weapons, designed only to disorientate the enemy.'

'And then what? Our boys pick them off one by one, or blow their arms and legs off with grenades to let them spend their final moments writhing in agony. Don't tell me that's better than what I'm doing. Chlorine gas is quick and almost painless.'

'Chlorine . . .' breathed Nernst.

'Listen to him, Fritz. Walther's not the only other chemist around this table,' Clara exploded. 'Before you destroyed my career and made a housewife of me, I was every bit as qualified as you. Chlorine burns the flesh.'

'Not in the correct concentration. It's quick. I've seen it work in the lab.'

'So that's how your suit trousers get covered in cat hair,' said Clara.

Nernst pulled a horrified face. 'No, no, no. The wind will dilute it. You can't keep it concentrated for long enough.'

Einstein could contain himself no longer. 'And illegal. Strictly against international law.'

'International law only applies if the *status quo* holds, and that is expressly what we are working towards changing,' crooned Haber, as calmly as if he were discussing stationery orders for the department. 'There will be a new world order.'

Nernst's brow was deeply furrowed. 'I cannot support you in the gas attacks.'

'I don't need your support.'

The older man looked momentarily stung but spoke indignantly. 'Victory will be ours by spring. There will be no need for chlorine gas. Remember, I've been to the front line, I've seen what it's like. Our lads are in the best of spirits and they're fighting hard.'

Rathenau drew attention to himself by elaborately lighting a cigarette. It smouldered between his fingers, a prop for him to wave around for emphasis as he addressed the room.

'A swift victory is, of course, what we all hope for,' he said deliberately, 'And none of us would doubt the courage of our soldiers, but we can speak plainly here, gentlemen. We are all friends. We all have German interests at heart.' Einstein said nothing. 'The British Navy has blockaded the Channel. That rules out Germany buying in supplies from America. We can get limited materials and food from Italy, but the fact remains – and make no mistake, it is a hard fact – that Germany grows only two-thirds of the food it consumes.' He swept his eyes across the leftovers, tureens half- full even though everyone had eaten heartily. 'We will starve if this war continues for years. The same is true for raw materials. Every rifle, every field gun, every helmet; all of them must be made from German metal and made on time. My job is to keep us in this fight as long as possible.' He flicked the grey column from the end of his cigarette into a crystal ashtray in one practised movement. The aroma of the tobacco teased Einstein's nostrils and made him wish he had remembered to bring along his pipe. He could do with a smoke.

Haber brought his fist down, setting the cutlery ringing. His eyes burned in the subdued lighting of the room. 'There! See! Thank you, Walther. You keep Germany in the fight and I will give us the means to end this war swiftly and easily. Revolutions are seldom bloodless. Our courage to do what needs doing must not fail us.'

Clara carefully stood at the other end of the table and began collecting the plates. She turned for the kitchen. Then she froze, lifted the crockery to eye level and dashed it all to the floor.

9

Ypres, Belgium

There were dead horses outside the cathedral. Their broad backs shone in the sunlight, giving a false impression of their being at rest, but it was the smattering of bomb debris and the splashes of dried blood on the flagstones that told the real story.

Lemaître bowed his head near one and prayed. Not for the horses but for . . . What was he praying for? The riders, now long buried? Or for himself?

He had stopped consciously wishing for the war to end sometime around Christmas. The conflict was a habit now, like humming the same tune over and over, or biting your nails. He looked down at his own ruined fingertips.

He never used to chew his nails. Now he did it without thinking, the same way he dropped to the ground at the sound of a rocket fizzing through the air. He'd learned to block out the sound of the big guns. There was nothing you could do against those. You were either underneath a falling shell or not, but the rockets and the rifles and the machine guns: those were personal. Someone had aimed the weapon at you.

Although it was spring, nothing much had changed. The German advance had stalled with the flooding of the polders, and both sides had dug in. The front line now stretched almost eight hundred kilometres from Nieuwpoort to Pfetterhouse on France's border with Switzerland, and most of it was made up of trenches, barbed wire, mud and corpses.

The draw of the cathedral was as powerful to him as a magnet to iron filings. Lemaître wanted to feel the caress of the light through the stained glass and listen to the cool quiet beneath the lofty rafters. Perhaps when they were dismissed he could steal back, light a candle and hope to smell the incense.

The light was fading fast. The men were heading for a rest, having spent the day marching up and down the line to the east of the town, moving from one skirmish to another looking for a place to be useful. As they marched through the town, curtains would twitch and townsfolk would sometimes watch. Mostly they were ignored, as if by turning a blind eye the remaining population could avoid the hostilities.

Lemaître pitied such naivety. A few days ago, out in the surrounding farmlands, there had been a woman hanging washing just a dozen yards from where Lemaître and the others had been pinned down by a barrage. As they cringed in the ditch the peasant simply adjusted her shawl and hung out her sheets.

Today, Lemaître had marched past the cottage. The washing was still on the line, scorched and ragged now. The dwelling was nothing but a smouldering shell and there was no sign of the woman. Lemaître willed himself to believe that she had been in the washhouse or somewhere else at the time of the destruction.

When he arrived at the wooden barracks built by the sappers, the cooks were filling the air with the aroma of supper. He closed his eyes and took a lingering sniff. The next thing he saw was the breathless commander jogging into the encampment.

'Up, lads! Something's brewing. The King needs us to be ready for further orders. We'll eat later.'

He always made it sound as if the monarch himself was waiting on the front line. The rallying effect was immediate.

The troops stood and shook the fatigue from their bones. They gathered their things and snaked back through the darkened city, towards the fight.

Out of the gloom came a terrible rumbling, and a frenzied mass of shadows appeared on the road ahead: cavalry.

Lemaître and the others flung themselves from the path moments before they would have been trampled. It was the French cavalry, tearing past in open retreat. No horse had just a single rider; two or three men clung to each charging mount. As the horses galloped, they raised clouds of choking dust that engulfed the Belgians cowering at the roadside.

Lemaître stared at the rout. Then he sneezed. At first he thought it was the dust, but the spasm drove a dagger into his brain and he winced at the sharpness of the pain. There was a pungent smell in the air that made him want to gag. His eyes began to prickle.

Something was terribly wrong.

Behind the frenzied riders came the infantry. They too were charging down the road, shrugging off backpacks and tunics, throwing away their rifles – anything to run faster.

Lemaitre gazed through watering eyes at their dark skins. He had heard that there was an Algerian regiment in the fight, but he had never seen anyone from Africa before.

'Cowards,' sneered the man behind him.

One of the Africans ran blindly off the road, careering headlong into the soldiers in front of Lemaître. The men he collided with scattered as though a grenade had landed in their midst, leaving him open to view. He was on his knees, eyes bulging and frothing at the mouth, which was frozen in a silent scream.

Other runners were dropping from the line too, clawing at their eyes and throats, stumbling in the road like wounded pheasants and bringing others crashing down on top of them.

'It's poison gas!' cried someone.

The Belgians panicked. They too were about to become part of the retreat when the commander called out, steely calm, 'Steady now, lads.' He led them away from the road and into a meadow, where they trampled on the early crop buds. 'We wait for orders,' he said, and sent a runner.

The men looked to each other for courage.

A dispatch cyclist handed over a message. The commander snatched it and struck a match. The long shadows this threw across his face lengthened as he read, and Lemaître was sure some of the colour drained from the man's face.

The commander called the men around him and addressed them with an air of normality. 'I know this will come as a disappointment to you, but we're to return to the town and regroup. The Germans drove a hole in the line using, as many of you have guessed, some kind of poison gas.'

The men gasped, not about the gas but the hole in the line.

Lemaître wondered why they were not being sent to defend it.

'I'm pleased to say the enemy have not made good their advantage. The Canadians have filled the gap and they're holding. The British are already on the march to come to our aid. We are to return and prepare for an early start. New orders will arrive at dawn.'

The march back was more a laboured plod. There was no attempt at high spirits. When the men reached the outskirts of town they stared incredulously. It seemed that every ambulance and every medic the army possessed had gathered to tend the wounded. The soldiers picked their way through hundreds of victims, maybe thousands. The wounded covered the ground, filling side-streets and intersections, rolling and groaning. Some smoked weedy cigarettes; others cried, their tears clearing lines down their grimy cheeks.

The groans and the agony were as constant as the distant rumble of gunfire and soon became just as easy to ignore. However, Lemaître could not shut out the small whimpers that escaped those who were in despair. He wanted to stay and help, but his body ached everywhere and his head still thumped from the whiff of the gas.

He turned and trudged away with the others, walking through the camp until he found a small knoll. There he had a good view of the eastern horizon and he crouched down to watch in the direction of the enemy. He ate a tasteless meal and discovered that no amount of praying could help him that night. Instead he just waited beneath the silent stars, knees hunched up before him, smoking everything he had in his top pocket.

His eyes swept the black horizon for silver strands of gas.

10
Berlin

The only way Einstein could deal with the city and its inhabitants was to pretend they did not exist. He hurried from place to place, sidestepping the queues outside food shops and taking detours to avoid the parks he had once strolled in for inspiration. Now they were home to show-trenches that demonstrated how cosy it was on the battlefield.

He had caught sight of ladies with their spring bonnets bobbing along at ground level, as if on a carnival ride, and the grinning soldiers guiding them along, all polished buttons and chivalry as they proffered their hands to help the ladies climb the ladders. The sight had been sufficient to engulf him in depression for days. So now he stuck to the roads, back streets if possible, where he could not accidentally overhear the triumphalist war talk.

Yet there was no way of avoiding the soldiers outside the university that afternoon. As he arrived they were milling around a car with its bonnet up. The uniformed driver was leaning in, winged gloves twisting as he made adjustments.

Einstein was about to lower his gaze and hurry past when he saw officers in smart uniforms appear from the university buildings.

One of them looked like – no, one of them was – Fritz Haber. Einstein stared. The chemist was wearing an officer's uniform, his monocle hanging from a black cord around his neck. He carried a pair of leather gloves in his hand, but his body language was at odds with this image of studied nonchalance. His jaw was set, and he walked self-consciously, as if someone had over-starched the uniform.

Their eyes met. Haber gave the briefest of nods and ducked into the car. The officers followed and the soldiers jumped on the wooden runners. The driver closed the bonnet, revved the car once, and drove it smoothly away.

Inside the building, Einstein found Nernst standing beside the corridor window, head in hands.

'Walther?' Einstein rushed over.

The chemist straightened himself and looked away, pulling out a handkerchief and sniffing loudly, trying to compose himself.

'What's wrong?'

Nernst looked up through bloodshot eyes. 'It's Clara.'

'Is she ill?'

Nernst shook his head. 'She's dead.'

Einstein stopped himself from saying anything trite and gradually coaxed the story out of Nernst.

Haber had returned from Belgium, flushed with the success of the gas attacks. Five thousand Allied soldiers had been extinguished without a shot being fired, and Haber had immediately been recruited at the rank of captain. During a celebratory dinner party he and Clara had fallen into a terrible row, much worse than the one before Christmas. Haber flatly refused to give up the gas research; Clara refused to accept it as anything other than murder.

That night, Clara took Haber's service revolver from his uniform and crept into the garden. There, in the darkness of the night, she turned the gun on herself and pulled the trigger.

The crack of the gun woke their son, who found his mother on the grass, bleeding from the chest. He had time to lift her head on to his lap before her eyes went as dark as the sky above.

Einstein shook his head in disbelief. 'But I just saw Fritz outside, getting into an army car.'

Nernst grabbed his arm; the chemist's whole body was trembling. 'That's the worst of it. He's left for the front again. They're to carry out more gas attacks, eastern front this time.'

'So he doesn't know about Clara?'

'Oh, he knows.' Tears sprang to Nernst's eyes. 'He knows she's dead, but he doesn't care.'

The shock of Clara's death and Haber's disregard for it was like tuberculosis that had taken root in Einstein's bones. As the sun dropped below the horizon and the day's lean promise of warmth collapsed, he made his way to Elsa's.

Numbly, he made a fire and sat looking at the crackling glow. Elsa cleared away the dinner things and came back through. She peered into the half-empty coal scuttle.

'I'll buy you more coal, Elsa,' he said patiently, almost kindly.

She settled in her chair and set about knitting a new dish-cloth.

Ilse glided into the room, preening herself as she did so. She was a taller, slimmer version of her mother. Her thick dark hair was squared off at the jaw and swept back from her face in precisely defined twists.

'You'll catch your death of cold,' said Einstein, looking at the calf-length summer dress she was wearing.

'No, I won't.' Her voice was full of sparkling certainty. She lifted her chin for a final inspection in the mirror over the mantelpiece. 'And if I do get cold, Georg will lend me his jacket.' She winked at her younger sister, who was curled up so tightly on the sofa it was easy to miss her.

'Tell me all about it when you get home?' asked her sister.

'I will. Don't wait up, Mama.'

'Back before ten, young lady.'

'Of course.' Ilse skipped from the apartment, forgetting to close the front door.

'Margot, go and shut that,' said Elsa with a bemused grin.

'Georg Nicolai?' said Einstein.

'Yes, you remember him. He wrote that paper no one would sign.'

'Of course, I remember him. I'm just surprised. I mean, he's so much older than Ilse.'

Elsa shrugged. 'He makes her happy. That's all I care about.'

As the evening wore on, Einstein could see Elsa eyeing the clock. Usually he had made his excuses to leave by now, so that he could calculate for a few hours before falling into bed. But he had no stomach for mathematics tonight. He cleared his throat. 'I wonder if I might stay here tonight, Elsa?'

Her face lit for a moment before her brow creased. She nodded towards the girls' room.

'I'll sleep on the settee.'

Her face lifted, though not so much out of happiness as contentment.

'I'd like you to stay,' she said, and bustled off to fetch the spare blankets.

Some weeks later, Clara's ashes were sitting on Haber's mantelpiece. A picture of her as a younger woman had been placed next to the ornate urn. She was holding a letter and looking through her eyelashes at someone out of shot. Einstein studied the picture. The more he looked at it, the more he thought how unhappy she looked.

Haber, back from the east, looked like a stranger in his own home. He wore a black suit and a similar countenance as he exchanged subdued greetings and handshakes with his guests. Einstein watched him, unable to fathom the tightly controlled body language.

As Haber passed by the fireplace he caught sight of the urn and stopped. He touched a finger to the gilded leafwork, and his chest heaved. The next moment, he turned and left the room.

'Wait here,' said Einstein to Elsa and set off in pursuit.

Haber was in the hallway, resting on the turned mahogany of the banister, nursing a drink.

'Would you like to talk?'

'It's not true that I didn't care about her,' Haber began. 'You know how infuriating wives can be, but I didn't want her dead. She screamed at me that night until I couldn't think straight. Nothing I could say would calm her. She just screamed and screamed at me until I was concerned for her heart. I had to walk away from her in the end. I didn't know what she was going to do.' Haber looked desolate; he glanced at the high ceiling and the magnificent cornice. 'Of course, I'll never be remembered for the fertiliser now, just the gas. Without me though, this country would have starved already.'

'Then give up the gas work. Turn your research back to good use.'

'I can't turn back now. The military are fully behind the work; I'm part of the chain of command. They're supplying more money, more staff. They want industrial-scale production: hundreds of men, maybe thousands, working to produce gas. And I'm to find better compounds . . .'

'Better?' Einstein spat out the word.

'Quicker,' murmured Haber.

They lapsed into silence.

'Would you stop if you could?' asked Einstein eventually.

Haber looked into his glass. 'I haven't forgotten that we promised you a physics institute to direct.' He drained the glass and hauled himself up the banister-post. 'I'd better be getting back.'

The reminder stung Einstein into silence, allowing Haber to return to the mourners.

Einstein and Elsa were half-way home when she turned to him. 'Will you stay tonight?'

Einstein shook his head. 'Not tonight.'

'But why not?' There was a note of pleading in her voice.

'Elsa, you know I relish the single life, and yet I have you as a dear little wife to return to whenever I need.'

'Whenever you need,' she said sarcastically. 'What about whenever I need? I've been many things to you Albert, but I am not your *wife*.'

'But you are, you are. It is how I think of you,' wheedled Einstein. A new thought struck him. 'Is this because we don't, er, *sleep* together? Is that what you want?'

They had been lovers briefly, but the wanting had turned out to be greater than the resolution. Their ardour had quickly dwindled. No matter what he thought about Mileva, she had always been so passionate.

'No, it most certainly is not, Albertle. Neither of us needs that complication. You've lied to me. There's no other way of putting it.'

They continued along the pavement, mingling with men toting briefcases, freshly discharged from the day's office work.

'Elsa, now is hardly the time or the place.'

'Then when is? When you decide it is? Why must it always be you who sets the agenda?'

'You know as well as I do that it is the man's role to dominate.'

She stepped squarely in front of him, forcing him to halt. 'Only in your male head. After Mileva left, you told me to have patience. Well, I've been patient for the better part of a year now, and for two years before that. How much . . .'

'I also told you to be content with what you had.'

A lamplighter brushed past them, lifting a long metal taper to the streetlamp overhead.

Elsa waited for him to complete his job. 'You have wormed your way into the girls' affections. Ilse thinks of you virtually as a father, Margot even more so. When I think of the letters you used to write to me, all about your undying love for me, the depth of your desire to be with me . . . I believed it all, and I've put up with this for long enough.'

They stood in the yellow glow of the gaslight.

'But the little bit of distance that we have from each other is a good thing. It insulates us from the banality of everyday life, the dullness of routine. It keeps us in love.'

'Do I look as if I'm in love with you right now?'

He searched her eyes for some warmth.

'I'm serious, Albertle. You're dallying with me, and I have two girls to find husbands for. The chances of me doing that successfully are improved by one of two things: either I become a married woman again, or . . .'

Einstein swallowed, waiting for her to finish.

'Or, I become single again. But I cannot and will not remain a woman who is having an affair with a married man.'

'But I promised Mileva I wouldn't divorce her.'

Thunder rolled across Elsa's face. 'Find a solution, Albertle. It's what they say you're good at.'

The pencil slipped from Einstein's fingers as a strange quivering took possession of his insides. He jumped up and pressed one hand to his heaving chest. With the other, he gripped the pad of writing paper. Nothing could force him to let go of that after what he had just written.

The morning light filling his study suddenly burned his eyes, and his balance began to fail. Even though the situation was terrifying, he wanted to laugh at the euphoria inside him. He steadied himself on the edge of the bookcase. How ironic, he thought, fighting breathlessness, to die from the shock of finally succeeding.

When his vision cleared, his eyes were looking at the portrait of Newton hanging on the wall. Could this have been how he felt?

Einstein collapsed back down at his desk. Bringing his breathing under control, he scanned the pencil marks on the page. It was still there, in his own handwriting, the number he had dreamed of deriving.

Beneath the tower of equations sat the answer: forty-three arc- seconds. It was the exact deviation in Mercury's orbit that the astronomers measured: the forty-three arc-seconds that were completely impossible to understand using Newton's law of gravity.

Newton! The greatest scientist who had ever lived. His work on gravity had led to the Age of the Enlightenment and changed everything. His way of analysing forces had bred the engineers that were now transforming the world with bridges, dams and skyscrapers. And now – Einstein grew breathless again – now, more than two hundred years of Newtonian thinking had been overthrown by forty-three seconds of arc.

There could no longer be any uncertainty. The universe was an invisible landscape of contours. General relativity worked.

No one could doubt him now.

PART II
Time

11

Berlin

1916

There was an uncomfortable silence when Einstein finished his talk. He looked out over the assembled fellows of the Academy. The light from the chandelier glinted from their monocles and the gilt edging of the portraits that lined the walls. Perhaps he had not explained himself correctly. He knew he had a tendency to celebrate the details at the expense of the basics. He opened his mouth again.

'Newton believed that gravity was a physical force carried somehow through space. He believed that space and time were absolute, a rigid framework that objects moved through. If you take away the objects then space would remain as an empty box. Relativity shows this is not true, that space and time only have meaning in order to explain the relationship between objects and events.'

He swept his arm around the room. 'Think of the volume of this room: what does that volume actually tell us? We may think of the volume as a measure of the space in the room but if you take away the walls, the room's volume ceases to have meaning because there is no longer a room – there are no boundaries from which to measure. So, really, the volume of this room is telling us about the placement of the walls relative to one another, and to the ceiling and floor.'

There were a few stilted nods.

'Everything must be measured relative to something else, or it has no meaning. A tram does not travel at thirty kilometres per hour; it travels at thirty kilometres an hour relative to the surface of the Earth – which is itself orbiting through space relative to the central sun. By thinking in this way, we can gain a more accurate understanding of Nature. I know it cuts against the grain, I understand that – I do – but we must not let that prevent us from daring to think differently about these things. Thank you.'

This time there was a smattering of polite applause. In the pregnant hush that followed, the astronomer Hugo von Seeliger placed his hands on his knees and heaved himself upright.

Einstein half-expected him to pull out his gold-chained pocket watch and make a show of checking the time, just to rub in the fact that Einstein had overrun.

Seeliger's bearlike head turned from side to side, taking in the audience. He stroked his close-cropped grey beard and sniffed loudly. For a moment it seemed as if that was the only comment he intended to make, then he announced: 'I am the President of the Astronomische Gesellschaft and I think I speak for all members . . .'

Einstein pursed his lips. Such theatrics were unnecessary, but the relish on the Academy members' faces was clear. They were expecting something memorable. Einstein made the smallest of movements to check that he still had the letter secreted in his jacket pocket. Anticipating something like this, he had picked up the note just before leaving the apartment.

Seeliger spoke with a measured delivery. '. . . when I say that I am deeply distrustful of relativity. The work of Isaac Newton has guided astronomers for centuries – and so far has not let us down. Yet, you would have us throw it all away because of a trifling error in Mercury's orbit.'

'Not throw it away, but I would have you recognise it for what it is: an approximation of the truth.'

The room filled with murmuring and chair-scraping. Seeliger widened his eyes as if Einstein had just blurted out a profanity.

Images of Kepler, the great German astronomer, filled Einstein's mind as he protested. 'We would not be standing here today discussing this if it were not for the eight arc-minutes' discrepancy in Mars's orbit. It allowed Kepler to discover the elliptical orbit where everyone before had assumed circular orbits for the planets. Lord Kelvin of Britain once said that all that remained – *all* that remained – for science was to make better and more precise measurements, as if the work by the theoreticians were complete and there was no more fundamental physics to be discovered. But I ask you . . .' He swept his gaze around the room. 'How else can science progress? It is precisely by better measurements that we will find the gateways to new knowledge and deeper understanding, because we will identify the gaps in our theories.'

That pleased the crowd. They voiced their derision for the Scot, and Einstein knew he was winning points, even if he had resorted to their nationalism.

Seeliger squared his giant chest. 'Herr Doctor Einstein, you said in your presentation that the deflection of starlight is the most distinctive feature of your theory. Not true. Newton's theory also predicts a deflection. There is no novelty in what you propose.'

Einstein leaned on the lectern. 'But relativity predicts twice the deflection that Newton does. So, the path before us is clear: let us be scientists, let us measure the deflection and lay the matter to rest.'

'A waste of time and effort. Modern hypotheses are not needed here.' Seeliger appealed directly to the audience. 'I have already calculated how dust in Mercury's orbital plane can affect its motion.'

'You suggest enough dust to block out the sun,' said Einstein to the man's back. 'I know astronomers who are ready to help. All they need is backing.'

'Astronomers, you say. Truly astronomers or merely observatory assistants?'

Einstein stood firm under the audience's scrutiny; clearly many in the room knew this was a jibe against Freundlich.

Seeliger continued: 'No serious astronomer will work on your ideas. They are unnecessarily complicated. We do not need to retire Newton just yet . . .' He tipped his head towards the audience. '. . . even if he was an Englishman.'

The room erupted in laughter.

Einstein reached into his jacket to retrieve the letter. 'Professor von Seeliger, you are familiar with the work of the astronomer Karl Schwarzschild, I believe?'

'Of course, he was one of my Ph.D students.'

'I have here a letter from him. He's taken great interest in my work on gravitation from the outset and he's seen my final formulation. He agrees with my calculation . . .'

Seeliger lifted his chin and played to the crowd once more. 'Then perhaps it should have been our good friend Karl – a man who is even now pursuing his patriotic duty against the Russians – who should be standing before us to explain *your* ideas.'

Again the room erupted in laughter.

Einstein was still simmering from the insult when Planck sauntered over. The rest of the Academy were jostling out the room, eager for the sherry being served in the lobby.

'That looked bruising,' said Planck.

Einstein folded the letter and placed it back in his pocket. 'No more than I should have predicted.'

'Is it really as simple as taking the eclipse measurements?'

'Yes,' said Einstein emphatically. 'I have made a clear prediction, but von Seeliger behaves like a wildcat towards Erwin.'

'Freundlich?'

'Yes, they try to crush him because he's young and eager to help. They insist he works on nothing but more star charts. It's so old-fashioned. German astronomers will be left behind if they are not careful. And the cameras, they're still lost somewhere in Russia. Without them and Erwin, there's no way to test relativity and prove it to them.' He threw an angry glance at the retreating backs.

Planck's face looked pained. 'I'm so sorry you had to endure such a rough ride. That's not what you deserved, even if your ideas are difficult. I still can't grasp them.'

'Don't worry. Tonight was nothing compared to what I'm about to get in Switzerland.'

Planck's face filled with curiosity.

'I'm going to ask Mileva for a divorce.'

12

Zurich, Switzerland

Eduard squealed with delight and threw himself at his father. The uncertainty with which Mileva had opened the apartment door evaporated and she took Einstein's coat with unnatural haste. Only Hans Albert remained apart, regarding his father with a quizzical look.

Mileva looked slim as she reached up to the coat rack. She was just half the width of Elsa, and more like Ilse. Gone was the padding that had crept on after the boys' births, and her outline took him back to when they had first met. He had not appreciated her shape back then, just taken it for granted, and had derived more pleasure from the way her eyes smiled at him across the lecture hall, and the aphrodisiac quality of their discussions.

Mileva ushered them all to the dining-room, where a plate of scones and a pitcher of apple juice were waiting.

Movement was difficult with the smaller boy anchored to his legs, but Einstein turned it into a game, much to Eduard's delight. Once at the table, which was crammed in next to an upright piano, Einstein could not help but notice the full butter-dish. In Berlin, the rationing was so acute that turnips had been the only things available in quantity last winter.

'Tete, go easy,' Mileva chided her younger son. 'We don't want Papa to think I don't feed you.'

The little boy's cheeks were bulging like a cherub's, and his lips glistened with butter. His eyes were as bright as Mileva's used to be. Eduard munched away, grinning and swapping looks with his more reserved older brother.

Mileva watched them devotedly, her face soft. The resemblance to her younger self was so strong that Einstein remembered how they had danced together when the first relativity paper had been accepted. Rather than music, their laughter had provided accompaniment, and they drank until their heads spun. They had ended up on the floor together, breathing the hot scent of wine fumes over each other.

'Albert? Your son is talking to you.'

Einstein smiled sheepishly. 'Sorry. Please continue.'

'Mama sets me mathematics questions and I answer them in my notebook. She marks and comments on them for me,' said Hans Albert.

'He's a good mathematician,' said Mileva. 'He understands binomials fully.'

Einstein's felt a pang of loss.

'Are you not pleased?' prompted Mileva.

'Very pleased. Well done, Albert. You're a bright boy, and your mother is a good teacher.' The words sounded hollow.

'Perhaps you and I could have a similar notebook, Papa? You could send me questions.'

'Perhaps we could,' he said uncertainly.

'Your papa is very busy with his own studies.' Mileva reached for the teapot and chivvied them to proffer their half-empty cups.

At the end of the tea the boys slid from the table, leaving Einstein and his wife alone. Now would be the right time to bring up the divorce, the first time that they had been alone since his arrival. He opened his mouth to speak but different words emerged.

'Thank you for letting me see the boys,' he said.

'You're their father,' she said tightly.

'It's just that I sometimes fear for what they must think of me. Perhaps even what you must think of me.' He watched her closely.

She folded the napkins from the table. 'They miss you.'

'I know.' Suddenly desperate for escape, he said, 'I wondered if I may take Albert out this afternoon. I've arranged to show him an experiment at the university.'

It was not a lie; he had written to his old colleagues but he had expected to visit tomorrow, once the divorce business was cleared up. Now he had to get away and think. He could not look into Mileva's dark eyes any more.

She beamed at him. 'I'm sure he would love that. Albert?' she called.

The boy padded down the stone steps behind his father, their descent illuminated only by the dim lightbulbs that hung like bunting from the walls. The boy's footfalls grew uncertain and began to drop behind. 'Papa?'

Einstein turned and smiled; the university's basement did have a certain dungeon-like quality to it. 'Don't be afraid. We're lucky my friends have let me bring you here. I want to show you something wonderful about the universe.'

'Down here?'

'The equipment must be isolated from vibrations and disturbances. Down here is the best place. Come.'

Hans Albert resumed his descent. When he moved it was as if the length of his limbs was taking him by surprise. It was clear that his body was preparing to change into a man's. His face had developed, and his mother's blunt nose and soulful eyes had taken shape, but the emergence of such adult features seemed premature to Einstein; the boy was not yet twelve.

'Why are you looking at me like that?' asked Hans Albert.

'I was thinking about how handsome you are going to be when you're a man.'

Hans Albert shrugged off the comment as just another daft thing adults say.

In the basement the air was clammy and laced with a faint metallic tang. There was a single working light dangling from the poorly plastered ceiling, throwing out just enough illumination to reveal a wide metal tank, on top of which was a platform of mirrors and cylinders.

Einstein poked around between the cables and switches on a power board in the corner of the room and a faint hum rose into the air. The upright cylinder on the platform began to emit a perfect orange light through a small window set into it.

'Is that water?' Hans Albert reached towards the tank.

'Don't! It's mercury. Poisonous. The apparatus floats on it; helps to isolate it and allows it to rotate smoothly.'

Einstein stepped closer and placed his fingers on the board holding the apparatus. It glided round under his touch. He brought it to a stop and then crouched to look through another cylinder, this one on its side.

'This is a small telescope,' he explained before turning a small screw near one of the mirrors. Back and forth he went, making minute adjustments and checking his progress through the telescope. When he was satisfied he beckoned his son. 'Now, look through here and tell me what you see.'

Hans Albert stooped and peered. 'It's a pattern, rings of orange light.'

'Very good, very good. It's an interference pattern, produced when the two beams of sodium light are brought together and combined. That mirror in the middle of the bench is not a perfect mirror, it lets half the light through and deflects the other half off at right angles. So, it splits the light into two beams that follow different paths. Then it combines them again and feeds them into the telescope. Now! Here's the crucial part.'

He rotated the apparatus with his finger so that it turned through perhaps an eighth of a circle. 'Look again.'

Hans Albert repositioned himself. 'It looks just the same.'

'Exactly! No change whatsoever. Never disbelieve your eyes, son; science is nothing more than common sense and this . . .' He pointed to the experiment. '. . . is the most important observation for the last two centuries, because no change in the pattern means that the speed of light never changes. No matter how fast you're moving, the speed of light is always the same. It's the only thing in the whole universe that behaves like that. If two trains collide, each travelling at fifty kilometres per hour, they hit with a combined speed of a hundred kilometres an hour, but light is different. No matter how fast you're going, when you measure its speed, it's always the same.'

'But I'm not moving, I'm standing still.'

Einstein chuckled. 'The Earth is moving in its orbit around the sun. Don't tell me Galileo went through all of that with the Inquisition for nothing.' He ruffled the boy's hair to show he was joking. 'When we moved the equipment around, we changed the ways the rays move with respect to the Earth's motion, and still the pattern stayed the same.'

Hans Albert's face creased. 'I think I understand.'

Einstein stepped closer to his son. 'This is the experiment that I used for my first paper on relative motion. The one I wrote when you were just a year old.'

'When you lived with us.'

Einstein had not intended to stumble on to such delicate ground. 'I love your little letters to me,' he said. 'I sometimes fear that you no longer wish to write to me.'

The answer was not what he had hoped for.

'I do get angry with you at times, Papa. Eduard sometimes dreams that you are with us. He says he can hardly remember you being at home.'

'But I'll visit when I can. Look at me, I'm here now. I'm not a ghost.'

'Why do you send Mama such nasty letters?'

'What does she tell you about my letters?'

'Nothing,' said the boy defensively, 'but I hear her crying at night after they arrive. Why must you be so horrid to her?'

Silence reigned. When Einstein spoke, it was in a deliberately soft tone. 'When I'm alone, working in my flat – you remember my desk, don't you, you can picture me there? – I think of what I'm doing as being for you. I do it because I love you. We can all share in it.'

The boy looked confused.

'Yes,' said Einstein, 'it's my devotion to you that drives me to work harder, to show you what I'm capable of doing.'

'For Mama, too?' There was a crippling note of optimism in the boy's voice.

Einstein dropped his gaze, as if the words he needed were strewn across the flagstones. 'Son . . . Look, I can confide in you. You're almost an adult now; we have a bond, a relationship deeper than . . . Well, we don't need your Mama to be part of what we have. We're father and son.'

'But she talks about when you return and we are a family again.'

A rage erupted from Einstein. He kicked the tank, setting it ringing. Hans Albert jumped in alarm.

The sight brought Einstein to his senses. 'That will never happen. It's impossible.'

'It's not impossible!' shouted Hans Albert, tears beginning to flow. 'Only you make it impossible.' The boy turned and fled for the steep staircase, arms pistoning up and down. ' I don't want to see you again.'

'Wait!' called Einstein.

Mileva sensed the tension between them the minute she saw them. 'Something wrong?' she asked.

'No,' they said in unison.

So there is a bond, thought Einstein, he does understand.

Hans Albert disappeared to his room and Mileva led Einstein to the sitting-room. The place was comfortably untidy. His eyes stopped at their wedding picture, displayed on the dust-flecked mantelpiece as if it still held currency.

Mileva saw him looking, but before either of them could say anything a delicate melody of piano notes filled the air. Einstein looked towards the sound.

'It's Tete,' she whispered.

'I thought it was Albert who played.'

'He does, but Tete's overtaking him already. It's hard to tear him away from it these days.'

Einstein tiptoed to the doorway. Even though he was over-whelmed by the size of the instrument, the six-year-old radi-ated control. Now and then his eyes would flutter and his head would glide in some movement that mirrored the passage he was playing. There was a vivid clarity in his simple technique, no stumbling across the keys, and every note was given appropriate weight.

'He understands music,' Einstein whispered. 'Some people practise for years and never understand the subtleties of phrasing. He has it. My boy has it instinctively.'

Mileva smiled at him from the settee. 'He gets it from you.'

He returned her smile. 'Think how he would sound on a concert grand.' His mind filled with images of concert halls and standing ovations: Tete would be a grown man in black tails, taking his applause and sweeping his arm to the box at the side of the stage, where Einstein would be on his feet applauding. They would exchange respectful looks and then Einstein would turn to share the moment with Mileva.

Mileva? Elsa!

The fantasy shattered. He was once again in the shabby apartment with its rugs and cluttered shelves.

'I get to listen to him play every day,' said Mileva with wonder in her voice. She had moved from the settee, her voice low and intimate, her mouth close to his ear.

He jumped, forcing her to back away and disturb Eduard, whose crystal melody stopped short of its resolution. His small head jerked round, uncertainty written on his face.

Einstein felt breathless. He had nearly fallen into the trap. 'We have to talk,' he said gravely.

Mileva backed away.

'We have to bring this to an end.'

She clutched herself and shook, stammering, 'No. You promised me, Albert. No divorce.'

'Mileva, we must talk about this like adults. It's not for me . . .'

'Then who?' Her eyes were wide open.

'Elsa,' he said quietly.

'You don't need a divorce to be with her.'

'That's not the reason. It's for her eldest daughter. She's young, innocent, beautiful. There's a danger that gossips will ruin her chances of marriage because of her mother's association with me unless I can marry Elsa. You wouldn't want to harm an innocent girl . . .'

Mileva flushed crimson.

'I was confident that you would understand. This isn't for me. Elsa has agreed to be cited. As part of the settlement, I will guarantee you the money from my Nobel Prize.'

Mileva glowered. 'Oh please, Albert, spare me the Nobel Prize talk again.'

Through her venom he could see the way her eyes were glistening, and the tiny tremble in her lower lip. He pressed the advantage. 'Just agree and I'll be gone. We're finished. We have been since Berlin. Why prolong this any more? For you it's a matter of formality – you and the boys already live alone – but for Elsa and me it's a matter of importance.

There are gossips in Berlin. Those girls have done nothing. They should not be tarred by this.'

Mileva collapsed into the corner of the settee, breathing heavily. 'And what of our sons?'

'Nothing will change. I spoke about this to Albert today. Once Tete is old enough, I will speak to him as well.'

'You spoke to Albert about this before speaking to me?' She bit her bottom lip, fighting for control.

Einstein waited for her to acknowledge him again, eventually shuffling his feet to try to catch her attention. For her part, she continued to stare into the corner of the room. A solitary tear crossed her cheek.

After some time, Einstein whispered her name. 'Please?' he said.

She hugged herself more tightly, looking more like the third child of the family than the mother. Then she nodded tightly. 'You win,' she said, and squeezed her eyes tightly shut.

'Thank you.' Einstein turned to leave.

Hans Albert was waiting in the hallway. Standing rigidly, he would not look at his father but held the door open.

'Remember what I said, there are advantages to being my son.' The words generated no response. Einstein clapped his hands softly against the boy's shoulders: still nothing. He was as unmoving as the coat rack standing beside the door. Einstein stepped out on to the landing.

The door slammed behind him so violently that the draught ruffled his hair. It was only later that he realised he had not said goodbye to Eduard.

13

Berlin

As the curvature of space forces objects to move into orbits, so mountains force train tracks to curve around them. The swaying rhythm of a carriage could often lull Einstein into a pleasant doze, but not today. The encounter with Mileva and his sons was too raw to permit him any rest.

The train rattled over the points back into Germany and, after what seemed like an age, the fields outside gave way to the buildings and streets of Berlin. The train slowed and the other passengers began to gather their belongings.

Planck was waiting on the platform, arms clasped behind his back. 'How was it?'

Einstein hefted the valise he was carrying. It was not the weight of his clothes but the journals that provided the bulk. 'It was all right,' he lied. 'I must admit I hadn't expected a welcome party.'

Planck's brow creased. 'I wanted to tell you that Karl's back.'

'Schwarzschild?'

Planck nodded.

'This is good news, indeed,' said Einstein.

The older physicist shook his head. 'No, it's not.'

The smell of carbolic clotted the hospital air. Einstein wrinkled his nose, not at the smell of the soap, tarry though it was, but at the lingering odours it was trying to mask. He shuddered inwardly and forced himself to continue deeper into the claustrophobic hallways. Head turning from ward to ward, he was eventually accosted by the Matron, who blocked his path with two firmly planted feet.

'May I help you, sir?' Her tangled grey eyebrows were formidable.

'I'm looking for Karl Schwarzschild. He's been injured and brought here from the front.' He flicked her a hopeful smile but she remained impassive.

'This way.' She led him back the way he had come.

They paused to let an old man struggle by on two crutches. As the patient drew close, Einstein saw that he was not a geriatric, rather his face was drawn in pain and his stilted gait was the result of an artificial leg.

Einstein doffed his boater for a reason he could not quite fathom.

Schwarzschild was dozing, head partly hidden by a ballooned pillow. His legs looked like sticks under the thin blankets.

'I'm Albert Einstein.'

The astronomer's eyes opened in surprise. He smiled and tried to push himself into a sitting position but winced with the effort.

'Can I help?'

Schwarzschild shook his head. 'Best you don't touch.'

He wore a pair of striped pyjamas with a frayed collar. Beneath the limp material, his neck was covered with blisters, some of which had opened into sores. 'I'm sorry, I must look frightful. I do apologise.'

'I thought you had been injured.'

'Not by bullets. Not even the gas. Whatever it is, I caught it in the mud. Started at my feet. My mother used to tell me off for not drying my toes properly after a bath. Looks like I didn't learn.'

The attempt at humour snagged Einstein and set his emotions tumbling. He forced himself to smile, but it was a feeble effort. He distracted himself by retrieving a chair from another bed. 'Your letter gave me the greatest joy to read,' he said, sitting down.

The skeletal face brightened into a smile. 'It's a wonderful thing that you have done. I still find it miraculous that from giving space a shape instead of nothingness comes a conclusive explanation of the Mercury anomaly. What do you call this invisible landscape?'

'The space–time continuum.'

Schwarzschild nodded. 'There's something truly profound in your work. I'm glad that we finally have the chance to meet.'

'Me, too. I'm just sorry I can't shake your hand.'

They were bandaged except for the fingers.

'I have your calculations. They are more precious; I have the touch of your mind.'

'I'm astounded you had the time to work on this.'

'Oh, most of the time you're just waiting around.'

'But you found a solution so quickly. Ten equations simplified and solved, a perfect description of space–time curvature around a spherical celestial object. You are perhaps the first person to truly understand my work.'

Schwarzschild closed his eyes momentarily and swallowed with difficulty. 'It didn't seem quick to me. I might not have been so keen to start if I'd known how long it would take me.'

'Well, you persevered. Thank you. Until now, I have had only one true ally. Erwin Freundlich.'

Something changed in Schwarzschild's eyes.

'Do you know something I don't?' queried Einstein.

'It gives me no pleasure to say this, but there's a growing resentment of Erwin. He's not precise in his calculations. He makes mistakes and doesn't seem to think it matters. Struve is . . .'

'I know all about Struve's disapproval, and von Seeliger has a mighty temper, too. He refuses to accept what I have done. The light deflection is the greatest game now. Erwin thinks that we can measure it from Jupiter rather than needing a solar eclipse.'

Schwarzschild shook his head feebly. 'Jupiter's deflection will be too small to measure. You must know that; they're your equations.'

Einstein's bravado dropped away. His ideas were naked before this man. He felt very weary. 'The astronomers shun me. The physicists are too busy with their investigation of the atom. I'm a lone voice.'

'You have me.' Schwarzschild's voice was momentarily impassioned. It was followed by a difficult moment. The temporary nature of the astronomer's support was clear to both of them.

'Forgive me, Albert, I'm afraid that we will not be able to come easily to agreement over Erwin.' Schwarzschild's voice was a ghost of what it had been a moment ago.

'Maybe you're right, but what kind of a fellow would I be if I rebuffed him now? After all the hours over the years he has devoted to my work.'

'What do others say? Beyond Germany?' asked Schwarzschild.

'Lorentz has grasped it. I'm hoping to hear back from de Sitter, but we're isolated from England and America. No one there would dream of reading a German theory at the moment.'

'The war will pass.'

'But human memory will remain.'

'We had to fight, Albert. It was our right.'

Einstein forced himself to nod. Deny the war now, and it was tantamount to saying that Schwarzschild was dying in vain.

Schwarzschild continued. 'You're going to need allies for relativity, and if Germany can't offer them to you then you must look to other countries.'

'But where?'

'There is one man who will understand.'

'Who? Tell me his name. I will write to him at once.'

'Eddington at Cambridge.'

'Cambridge? England? We're at war. I can't just write to the enemy.'

'You'll find a way.'

Einstein noticed a pencil stub and a sheaf of papers resting on the white sheets. 'More calculations?'

Schwarzschild lifted his bandaged hands. 'I ask them to leave the pad there, so that I can imagine writing on it. Once I'm out of here, I'll bring the calculations round and talk you through them.'

Einstein dared not look into those sunken, dried-up eyes. 'I'd like that very much.'

There were too many funerals in Berlin these days. Schwarzschild's was the latest. All were conducted with the same stiff formality. Einstein looked around the congregation and wondered when the population would realise that this was the price of their ludicrous nationalism, that the quick victory they had been promised was impossible.

There was Planck, ramrod straight, belting out the hymns. No one would guess that his second son had been killed at Verdun and that his first was still listed as a prisoner of whom there had been no word for over a year.

Nernst was at Einstein's side, rumbling along to the verses. He, too, had received the awful message about one of his sons. His retaliation had been to swallow his previous objections and join Haber in the laboratory to search for deadlier compounds of gas.

The gas had not been decisive and had only driven the Allies to seek revenge with their own. Only the day before Einstein had seen a young man stumbling through the streets with a cane to compensate for his destroyed eyesight. He was clutching a tin of coins, having been moved on by the police for begging.

Still Germany's capacity for self-delusion towered. Deep down they must know their country's isolation meant defeat was inevitable, yet they chose to continue the charade and babble about victory.

The blossom was gone and summer was upon the city. Unable to resist, Einstein spent the morning pacing through the Tiergarten. Avoiding the show trenches, he stalked his thoughts from one avenue to another. He was wearing a comfortable, old, creased linen suit but had forgotten his hat today. He was grateful for the shade from the trees. Centuries ago, under the same leafy canopies, the royal family had come to hunt. Einstein hoped they had been more successful.

The true meaning of relativity pricked him. Proving the theory via a light-deflection test had become so important to him that it had become a fog in his mind. He had written to Lorentz in Holland to ask if he could put the theory before Eddington; perhaps the Englishman would listen as the request came from a neutral Dutchman. Now he had to put it aside and concentrate.

Schwarzschild's calculations had been so elegant, the product of a clarity of thought that Einstein seemed rarely able to muster these days. He knew there was more to relativity than just Mercury's orbit and the deflection of starlight. Sometimes he could almost see it, like glimpsing a moving shape through the fog, and he was sure it was profound, a way of using relativity to investigate not merely individual celestial objects but the universe as a whole.

But he could not yet find the way to see it clearly.

Eventually he gave up his wanderings and headed for home. He was crossing the tramlines of the Alexanderplatz when he spied a familiar shape sitting alone at a café table. Nernst was holding a cup and saucer, using each nervous sip as cover so that his bulging eyes could scan the groups of people coming and going across the concourse.

Einstein approached. 'You look troubled, Walther.'

He caught the chemist off guard. 'Sit down, Albert,' he said urgently. 'Tell me what you see.'

Einstein glanced around the other customers. People in their summer clothing were passing like clouds in the sky.

'Have you noticed?' asked Nernst, not waiting for a reply. 'All Germans together. No Jews, except us – and no one's talking to us. They still don't trust us despite everything we've done. Fought alongside them at every step.'

Einstein's shoulders dropped. Not this sad lament again.

'I thought it would be different,' Nernst continued, 'that we would stand and fight together, German and Jew side by side. I thought it was what my Rudolf really died for, not territory but acceptance.'

'Is that really how you see yourself? A German Jew?'

'What else? You should join us.'

Einstein spluttered, halfway between humour and derision.

Nernst's hair was unkempt, and his breathing was laboured even though he was sitting down. Despite the rationing, his pot belly seemed larger than ever.

'The German Jews need you. If your ideas really are as revolutionary as you profess, then declaring them as German will greatly help.'

'German Jews!' scolded Einstein. 'There is no such thing. In the eyes of the Germans you are still a Jew. I'm still a Jew. We're all still Jews. Barely tolerated.' A cold hand reached into Einstein. 'Oh my . . . Walther, is Gustav safe? Tell me nothing has happened to him as well.'

Nernst's eyes clouded.

'I'm sorry . . .' began Einstein.

'No, he's alive, but . . . but they've pulled him back from the line, put him on supply duty. Just him. None of the others, none of the other Germans. Then, a week or so later, they asked him to complete a questionnaire about his current duties. Not his service record, just his current duties.'

A dreadful realisation spread through Einstein. Now he understood why Nernst was upset. 'So, finally, German Command acknowledge that they cannot win.'

'I think so.' Nernst's face was bleak. 'They're looking for scapegoats.'

Einstein looked around again. Whereas before he saw strangers passing by or conversing, now he saw accusers, angry people willing to swallow the government line about why the war was lost: lazy Jews sapping morale and not pulling their weight. A man with a sharp face glanced over. Einstein was sure he was sneering and rapidly looked away.

'Germany is going to become a dangerous place,' hissed Nernst, resuming his surreptitious watch.

'What else can nomads expect but to be outsiders wherever we go?' said Einstein forlornly.

'Germany is my home.' Nernst jabbed the table, making his cup rattle and drawing looks of disapproval. He added in an undertone, 'It has to be, we Jews have no other country to call our own.'

Einstein fancied he could already see accusation in the eyes of those around them. Something hardened in his heart. 'Perhaps that's the root of the problem.'

14

Ypres

The dim gloom of the dugout and the peaty smell of the earth were things that Lemaître had learned to relish. Sheets of corrugated iron kept the walls secure and canvas sacking provided flooring on the duckboards. The shelter would offer no defence against a lucky hit during a bombardment but at least it kept them hidden from machine-guns and snipers, and rain clouds.

Wooden bunks were three high along the walls, leaving a central area for eating around an old card table. Lemaître was on a bottom bunk and, although he could not sit up, there was just enough space to hold a book at a comfortable reading distance. If he wanted to write, or work through a derivation from one of his textbooks to investigate how the author had arrived at a conclusion, he did so by lying on his side. The straw mattress did not provide the most stable writing surface, nor the most comfortable sleeping one, but it was adequate for both, and he was grateful.

'How does the war end?' asked Pierre, wearily rubbing his sunken cheeks. It was one of their verbal refuges, a way to start a conversation when no one knew what to talk about. It provoked a list of automatic comments that Lemaître listened to but seldom joined in until the real conversation began.

    — *Negotiation, surely. The return of Belgium.*
    — *No, a big push to drive them back into Germany.*
    — *What if it's they who make the big push?*
    — *What's stopping them now?*

The gas lamp on the table cast its earthy glow, warm in colour if not in its ability to heat their abode. Lemaître was counting the days to warmer weather, unable to shake the fear that tripping over his blanket could be the difference between life and death if a night attack came.

Thankfully, they were rare. Like any job, the war had settled into a mundane routine. Attacks mostly came in the grey mist of dawn. Once the danger had passed and the sunlight was fully over the land, both sides settled for breakfast. This was followed by the routine drag of repairing trenches, filling sandbags, standing-to with rifles aimed over no-man's-land in case the enemy tried a charge. For Lemaître and his squad, the day also included surveying and calculating bomb ranges. He had been transferred from infantry to artillery.

At least the soldiers were getting meat. The word was that in the cities protein was almost impossible to come by. There was a story doing the rounds of a packhorse that had collapsed, and no sooner had the soldiers redistributed its burden and ended its misery than the townsfolk had descended with knives and butchered the animal to the bone in minutes.

– *Germany is landlocked and growing weaker; they must collapse.*

– *There'll be a revolution inside Germany.*

– *What if the next lot's even worse?*

'I'm not sure I'll remember how to behave back on the streets,' said Pierre, taking the conversation off on tonight's tangent.

'Well, you can stop belching for a start. Have some manners.'

'There's always the chance of a promotion. Being an officer must help keep you safe. Don't you think, Georges?'

Lemaître glanced over from his book. 'I'm not cut out for promotion: wrong attitude.'

Smiling was something that hardly any of them did spontaneously any more; instead they had learned to recognise a kind of grimace as the substitute.

Michel made the gesture now. 'After what you did today, you're lucky you're not back up in the forward trench.'

The comment brought a round of tired but good-natured agreement.

Lemaître had pointed out to his commander a mistake in the artillery manual's mathematics. Without correction, the guns would never find their marks. The commander had not liked the interruption.

'I did what anyone else would have done,' said Lemaître. His story made him think of Jacques. From their correspondence it was now clear that his brother had taken to the military life and was rising up the ranks.

'I think I'll become a mechanic after the war,' said Louis. His bunk was adorned with sculptures he had created from spent artillery shells.

They were infinitely better than the souvenirs some of the other dugouts were collecting. Rats were a constant problem. Not content with just killing the vermin, some squads were hanging the carcases as trophies by their tails from wooden frameworks.

Pierre caught Lemaître's eye. 'What about you? Back to university?'

'To finish my thesis, yes. If there's anything left of the university. It was in Louvain.'

Mention of the sacked city threw a pall across the proceedings. They all knew people who had died in the wanton violence there.

'We'll be a generation of old men before we're even thirty,' said Louis.

'Why? We'll recover, once we get away from these stinking holes.'

117

'I don't mean like that. Physically, yes, we'll recover, but I mean in our attitude. The way we think about life . . . and death.'

'Now you're just being morbid.'

'No, I'm not.' Louis was a neat man whose once young face grew more gaunt by the day. 'All of us wake up and wonder if this is going to be our last day. We have no control over it. Sure, we can keep our heads down but we can't stop a shell from just dropping out of the sky. Can't even see it coming. This must be what it's like to be old. Waking up every day and wondering if it's your last. Perhaps it's confronting your own mortality that robs you of youth, rather than the years ticking by.'

'My belief in God and the knowledge that death isn't the end is the only thing that stops me going insane,' said Pierre with an embarrassed little laugh. He scanned the room, eager for confirmation.

'We'd all like to believe that.'

'Sounds a bit too convenient for my liking.'

'Your faith will fade once peace is upon us.'

'Georges, back me up here. You believe, don't you?'

Lemaître put down his book. 'I do believe, but faith is not about the afterlife. For me, it's deeper . . .'

'Deeper!' Michel leaned back from the table. 'What could be deeper than having a bullet smash through your heart?'

'It's difficult to explain. The natural world is beautiful, with so much order . . . Isaac Newton once wrote how it led one to believe there is a creator, that the whole thing had been designed. I don't believe that the design was fixed; there has to be change and evolution of landforms and animals, but the underlying principles of Nature are beautiful.'

'Exactly. God must be real to explain all the things that science can't. There's just too much we don't know.'

Lemaître softened his voice. 'I'm not sure you need God to do all the things that physics cannot explain *yet*, but I do believe you need him to make the laws of physics in the first place. I believe the Bible may even contain scientific knowledge, if you know how to interpret it.'

He could see the sceptical look on his comrades' faces.

'I don't mean that the ancients practised science, but think of Genesis, and the creation of Heaven and Earth. The immediate thing that God did was to create light, and the Earth. In 1905, millennia afterwards, a German . . .' He ignored the rude noises and words. '. . . called Albert Einstein was working out the consequences of motion and derived an equation that shows light and mass are connected – almost certainly interchangeable. Light can be transformed into solid matter, and vice versa. How could the ancient scribes have known that light and Earth went together without divine inspiration?'

Pierre looked doubtful, possibly even somewhat embarrassed. 'I believe in . . . something,' he said lamely.

'That's a start.' Lemaître forced himself to smile. 'After my Ph.D, I'm thinking of applying to a seminary.'

This was, he realised, the first time he had said it aloud. Until this moment, the thought had been known only to him and to his Maker. Now he had turned those evanescent feelings into words and set the air vibrating with them, casting them from the spirit world into the material one. He felt elated.

'A priest? I thought you were going to become a physicist?' asked Michel.

'Why can't I be both?' asked Lemaître, genuinely puzzled.

Pierre reached for the lamp and extinguished its flame. 'Another day tomorrow, lads, best get some rest.'

Outside they heard the tramp of boots on duckboards. It was probably a squad returning from barbed wire cutting in no-man's-land. They were the only ones who moved around much during the night. Lemaître thought of a story he had heard the previous morning. A cutting squad – perhaps the same one – had come virtually face to face with a German group doing the same thing to the Belgian defences. The two squads had stared at each other like polecats in a stand-off, then turned silently and retreated into the night.

Lemaître had thought about the encounter all day. Even in the muddy landscape of water-filled craters and bomb-blasted trees, where there was no nature or beauty for God to work through, He still found a way to operate. He was in the hearts of those patrolmen who chose to back away.

Lemaître laid aside his book and settled back. The straw mattress felt quite comfortable that night.

15
Berlin
1917

According to the law of entropy, all systems tend to move from order to chaos. It is an inevitable process, inescapable and irreversible. Einstein needed only to look at the mess in his office to know that it was true. Piles of journals had toppled over; unfolded letters littered the desk.

Everything could wait.

He had been right: general relativity could give him a way to describe the whole universe, not just the contours of space around an individual object, but the underlying curvature of the whole universe.

But it was a colt that needed breaking. Just when he thought he had succeeded and order prevailed, it would rear up and throw him again. Something in the maths would make no sense; the formulae would suddenly insist that the universe was like an expanded sheet of rubber, or a collapsing building.

It reminded him of something. He lifted his head and his eye caught the portrait of Newton.

Newton had always known that his gravitational theory was flawed. Since all the stars generate gravity, the philosopher had wondered how the universe could be stable and not collapsing. His solution was that the whole universe was the Sensorium of God, allowing the Old One to directly intervene and hold the universe in shape. Now Einstein faced the same problem. But he didn't need God to solve it for him. Einstein winked at the portrait.

Fighting to remain calm against the electrical surges the work was generating within him, he hunched over his writing-pad. The key was in the amount of matter the universe contained and the way it was distributed through space. He needed something that would act to oppose this. As he stared at the equations, a solution presented itself: there must be a kind of energy in the universe, unknown on Earth but capable of resisting gravity. He could see how it would fit into his equations quite simply, but the arbitrary nature of it bothered him.

What else could he do? One only had to look into the night sky to know the universe was static, the stars unchanging in their constellations for eternity.

He took his pen and recast the equations. When the time came to enter the new term he marked it as the inverted V of the Greek letter lambda. In his head, he called it the cosmological constant.

Einstein's days passed in frantic episodes of activity followed by indolence, removed from the normal run of space and time. It mattered little whether it was light or dark, warm or cold, mealtime or bedtime; if inspiration was upon him, he would work.

When it was clement, he kept the windows open, hoping that the breeze would dust the study. As the crisper weather arrived, he would look out at the evening stars, breath escaping from him in billows of vapour, and ponder how ironic it was that he could experience more of a connection to the universe through his mathematics than by standing under the night sky.

Things were difficult with Elsa at the moment; the divorce negotiations with Mileva were dragging on and his cousin was convinced he was stalling. When the urge to see her and the girls did hit him, he would have to remember to check the clock as he reached for his hat, to make sure that it was not the middle of the night. If they were not at home, he would continue to the university.

He saw Ilse more regularly these days. At his suggestion she had become the secretary to the physics institute he headed, the one for which Haber had scraped together the funding. Not that it was much of an institute yet, just Ilse and him, and Erwin Freundlich, whom he had managed to rescue from the old naysayers at the observatory.

He was at the university, staring in disbelief at a letter – thankfully not about the divorce – when there was a knock. Freundlich stepped in and closed the door behind him.

'You wanted to see me?'

'Come in,' said Einstein, still preoccupied with the contents of the letter. 'Erwin, I need your advice. Is there any evidence that the universe might be – this is going to sound a little strange – collapsing?'

Freundlich's eyebrows rose.

'Or expanding?'

'Expanding?'

'Yes,' said Einstein, 'the space between the stars, perpetually driven to get bigger or to shrink. It seems nonsensical to me when the constellations have persisted for aeons.'

'Well, the stars are in motion but very small, still difficult to measure and apparently quite random. They would almost certainly average out to zero, I think.'

Einstein nodded emphatically. 'And these gas clouds I hear about, the nebulae, are they the same?'

'I think so. Although I have heard that some people think the spiral-shaped ones are vast collections of stars, like our own Milky Way but much further away. Is this about the cosmological constant?"

Einstein waved the letter. 'De Sitter doesn't like it; tells me I'm plainly wrong. Seems everyone thinks they know more about my theory than I do. He thinks it unnatural and says he can find a solution to the equations that doesn't need it.'

'De Sitter? In Holland?'

'The trouble is he has to leave out all the matter in the universe to make the equations balance. How can that be right?'

Freundlich spoke a little more loudly. 'You're talking about relativity with other astronomers?'

'I'm sorry, but I have to. We still have to prove relativity.'

The astronomer looked puzzled. 'Then get me back the confiscated equipment. We can be ready as soon as the war ends – sooner if we can find safe passage. There's an eclipse next year across America, remember? We don't need help. We just need the equipment. I can do this, Albert.'

Einstein retreated behind his desk. 'Erwin, the equipment is lost until this war is over.'

He knew that what he had to say next was not going to be easy for Freundlich to hear. 'An English astronomer called Arthur Eddington is publishing a series of papers about relativity. They're being written by de Sitter; Lorentz has organised the whole thing at my request. He is also forwarding my original papers, though they can scarcely be published at the moment. Eddington understands and is helping to plant the seeds of the theory in England.' He drew breath. 'The papers will emphasise the eclipse across America in 1918 as a way of testing whether relativity has any value.'

Freundlich resembled a candle, melting beneath a flame.

'You will have your chance, Erwin, but right now we need allies.'

The younger man lifted his gaze from the carpet, nostrils flaring. 'You've killed my career! What chance do I have now to prove myself?' He turned and blundered across the room.

Einstein rushed out from behind the desk, heart thumping. 'Erwin, please . . .'

A sharp sensation in his stomach brought him to a halt. Sweat broke out across his brow. He doubled over in pain and fell heavily against a bookcase.

Freundlich tried to catch him but the physicist crashed to the ground. 'Albert?'

Einstein spoke through gritted teeth. 'This is a bad one.'

'What is?'

'I felt it in Switzerland . . . with my boys. I think, I think I have a cancer in me. Getting worse. Take me home.'

'I'll take you to the hospital.'

'No, please, just take me home.' He had no strength to beg.

Freundlich struggled to get Einstein to his feet. The exertion was almost more than Einstein could bear. Everything seemed to come and go in snatches around him, but there was one sound he would never forget. When Freundlich opened the office door, Ilse had screamed.

Elsa was in a flat spin, tidying his apartment furiously under cover of the doctor's visit. From his bed, Einstein listened to her chaotic progress through his home, oblivious to the movement of the stethoscope across his chest. A nearby door creaked open, followed by silence. That could only mean one thing. Doctor or no doctor, Einstein could contain himself no longer.

'Not my study, Elsa! Tidy anything else you want, but not my study.' He could picture her standing there, blowing through pursed lips as she surveyed the mountains of papers and books, assessing her best route into the terrain. He began to kick the covers from his legs.

Elsa appeared just in time to stop the reluctant patient from tumbling to the floor.

'It hurts just to move,' Einstein complained.

'That is precisely why I have asked you to lie still.' There was no mistaking the annoyance in the doctor's voice. He stood on the far side of the bed, stethoscope hung round his neck, eyes narrowed.

Einstein wondered again where Elsa had found the man.

'I'm glad you're here, Frau Löwenthal. It saves me explaining twice. I'm now certain that Herr Einstein does not have cancer. He has a swollen liver and, I suspect, a gastric ulcer. He's in no danger of dying, but he will be an invalid for many weeks, possibly months.'

'Months?'

'Yes, months.' The doctor looked as if he was enjoying giving the diagnosis. 'You've been overdoing it.' He forestalled any riposte by turning to pack his bag. Einstein muttered his thanks, embarrassed at ever having voiced his own dramatic diagnosis.

He listened irritably to Elsa showing the doctor out. When she returned he said, 'I must tell the boys. They must know their father is ill.'

'And what good will that do, except worry them? Especially when their mother is poorly, too.'

Letters had arrived from friends in Zurich and from Hans Albert informing them that Mileva had been hospitalised with chest pains. The boys were being cared for by a maid.

'Mileva's not ill. I've told you before, she's a prima donna, always writing her mood in the darkest colours available. It's all a ploy to delay the divorce.'

'She's been diagnosed by doctors. I'm not one to take that woman's side, but you must be reasonable.'

'I will look after my boys, if she really is ill ... if she dies ...'

Elsa's wide mouth gaped. 'How can you be so callous about such a thing? That's the mother of your children you're talking about.'

'Death is an inevitability. Why be perturbed by it?'

'Oh, you say that too often. It's one of your lines. All a pretence! Anyway, you can't even look after yourself. How do you plan to look after your sons? If it weren't for me, you'd be in a sanatorium.' Without a backward glance, she strode out of the room.

Einstein seethed. He had a nagging feeling she was not too unhappy about his infirmity.

He developed the habit of pretending recovery by sitting in an armchair by the window. He would have preferred birdsong to the sound of the spluttering cars, but the parks were too far away for him to walk. From his chair he read the letters documenting Mileva's recovery with a grim satisfaction.

He was interrupted by the sound of Elsa's key in his apartment door. It took all his strength to get to the hallway to greet her with some fake vivacity. He then sat back down and tried to mask his exhaustion by picking up a book and pretending to read it. In reality he was having trouble bringing the type into focus.

Elsa unpinned her hat and looked at him with a smug expression. 'Guess what? Frau Keller is moving away.'

'Really?' he said slyly. 'I don't know who you mean.'

'Yes you do, we've talked about her. She lives in the apartment opposite mine.'

Einstein lifted the book. 'Nernst gave me this, said it reminded him of me.' He angled the spine for Elsa to read. She hardly even glanced at it, distracted by a damp towel he had left in a heap on the night table.

'*Tycho Brahe's Path to God* by Max Brod,' supplied Einstein. 'I met the author in Prague, where I was giving a lecture about special relativity. He was in the audience.' She turned her back. 'Tycho died in Prague. He compiled the star chart that Kepler used to find the laws of planetary motion.

Nernst thinks that Brod has written Kepler using my characteristics.'

'Does he constantly deflect the conversation when it doesn't go the way he wants, too?' Elsa folded the towel and patted it a little more heavily than was necessary.

With a sinking feeling Einstein lowered the book. 'What's wrong with the present arrangement?'

'You're not the one carrying stewpots down the street, having to deal with all the beggars staring at you because they can smell sausages. You don't have to avoid their hungry faces . . .'

'I thought you liked the food. What's the point of my having connections if I don't use them?'

'I'm not saying don't use them. We're all grateful for the food you provide, especially the girls. But when there are no sausages in the shops, and I'm walking around with a stewpot reeking of them, all I'm saying is that it's clear they've come from the black market.'

'There're not black market; they're from friends. You could cook here.'

'And then I'd have to carry the girls' portions back home. Apart from all that, it makes sense to be nearer. Especially since we'll soon be married.'

She moved on to tidying the blankets.

Einstein's mouth was dry. He needed water. Nothing had been decided about the wedding. They were still exchanging letters with Mileva's lawyer over the exact settlement.

'But I still don't see why we have to move so close together. Couldn't we find a place in the next block, rather than just across the landing?' he said.

'Albertle! You'll move to Frau Keller's apartment or I'll stop cooking your meals.'

He stared dumbly.

'Agreed?' she prompted.

He nodded meekly.

Bounding across the room, she planted a maternal kiss on his forehead. 'Good, that's decided then. I'll make the arrangements. Now, can I get you anything?'

'A large glass of water, please,' he croaked.

She brought him the drink and he picked up the book again. He tried to lose himself in the story of Kepler and his personal struggles with the great Tycho Brahe. Each had needed the other – Tycho needed Kepler's maths, Kepler needed Tycho's observations – yet they could not agree on what to believe. Kepler believed that the sun was the centre of the universe; Tycho clung steadfastly, incorrectly as it had turned out, to the Earth being the centre of everything.

Einstein found himself musing that if he were cast as Kepler, as Nernst insisted, who was his essential nemesis, his Tycho? There seemed to be so many who could fill the role.

16

Cambridge, England
1918

Once Arthur Eddington had arranged himself in a seated position, he tried not to fidget. As in all things, it seemed to him best to get it right the first time. It was a tactic that also meant he was less likely to draw attention to himself.

He sat on the wooden bench in the long corridor, one leg hooked over the other, his back straight and his hands primly in his lap. His companion – Frank Dyson, the Astronomer Royal – suffered no such qualms about readjustment, pulling at a dense black moustache, raking a fingernail through the natural arch of his eyebrows, and drumming his feet on the black-and-white tiled floor, sending echoes up and down the corridor.

'How can you just sit there? Aren't you worried?'

'Conscience is the opposite of worry. One worries only when one is unsure of one's own mind.' Eddington spoke as precisely as he sat, all traces of his Kendal accent having been erased long ago, first by his schoolmasters, then by his own careful design.

'I don't mean about your decision, I mean about what they could do to you.' Dyson nodded in the direction of the heavy wooden door that led to the tribunal. 'You can go to prison for refusing to fight. Or they'll send you to the front regardless.'

'In which case, I'll carry stretchers. The war is coming to an end. The papers are full of the advances being made across the Somme.'

'I'm somewhat reluctant to believe the papers these days. Seems we've been close to victory since the whole bloody thing started four years ago.'

'Even if the Germans counter-attack, our case for exemption is strong.'

'Let's hope they see that. I must admit that I'm not sure I truly believe it.' Dyson lifted his eyes to the ceiling.

'Einstein's work is right, mark my words.'

'How can you know that so certainly without doing the test?'

Eddington looked at his companion. 'Because the mathematics is so elegant.'

'Don't look at me like that.'

'Like what?' asked Eddington.

'Superior, with a smirk.'

'I'm not smirking.'

'Yes, you are. You do it all the time. Most off-putting to strangers.'

Eddington considered the criticism for some moments. 'Perhaps that's what my face looks like normally.'

A door opened, disgorging a pompous-looking official who was looking down his nose at his notes. 'Arthur Eddington?'

Both men stood; Eddington nodded.

'Just Mr Eddington.'

Dyson took a small step backwards. 'It's *Dr* Eddington.'

The man looked unimpressed and indicated the open door.

Dyson muttered nervously in Eddington's ear. 'One more thing: I may have forgotten to mention that Mr Einstein is a German.'

Eddington turned slightly and nodded. When he spoke, his plummy voice was lowered so that only Dyson could hear. 'A wise precaution. And yes, I *am* as worried as you. I just choose not to show it.'

The tribunal consisted of three aged men, wrinkled as prunes, with heavy-lidded eyes. Eddington positioned himself on the spindly wooden chair before their heavy table. The man on the right wore a suit with dandruff-covered shoulders, the one on the left had frayed lapels and was barely awake, and the chairman looked over a set of notes with studied indifference.

'You are?' said the chairman.

'Arthur Stanley Eddington.'

'Born?'

'The twenty-eighth of December 1882, in Kendal.'

The chairman rolled his eyes in Eddington's direction. 'Kendal, in England, I presume?'

'Yes, in England.' It was hard to keep the sarcasm out of his voice.

'You are thirty-five, unwed, passed your medical with grade two, yet for the last four years you have been exempt from active service because your work at the University of Cambridge's Observatory has been deemed,' a slightly disbelieving tone entered his voice, 'to be in the national interest.'

'With respect, sir, I am Plumian Professor of Astronomy and Director of the Cambridge Observatory.'

'Quite so. The exemption ends on the first of August this year, and you have applied for a further exemption on religious grounds.'

'I am a Quaker, sir. We are pacifists.'

'That won't matter to the Kaiser and his hordes.'

'Maybe not, but it matters to me.'

The chairman tutted. 'And what would happen if we all felt like that?'

Eddington wanted to say that if everyone was a pacifist there would be no wars, but he sensed the chairman was talking only about England. Eddington found himself wondering if he really did look smug. He decided to say nothing and wait for the chairman to speak again.

'I have here a letter of support for your exemption from a Dr Frank Dyson, Astronomer Royal. He says that your research in astronomy should be ranked as highly as Darwin's in the zoological sciences. And that there is an eclipse in May next year that you are preparing to observe in order to . . .'

'To weigh a beam of light, sir.' Eddington jumped in with his prepared line. He knew he did not have charisma on his side and over the years he had come to rely on manners and erudition. He had also learned that the occasional quip could go down well. Yet he sensed that today was not the day to deploy his analytical humour.

'Weigh light?' said the chairman.

The two gargoyles either side of him stirred.

One said, 'Something we can tax perhaps?'

The chairman turned a sour eye, confirming Eddington's assessment about humour being inappropriate.

'Why the sudden urgency, Mr Eddington?' asked the chairman. 'I'm given to believe that these eclipses come around every year or so.'

'They do, but this one is particularly important for the testing of general relativity.'

'And that is?'

'The new theory of how gravity is generated by mass.' Eddington ploughed on. 'This eclipse occurs in front of the Hyades star cluster, a particularly rich grouping of stars, which will allow the deflection of starlight to be measured much more accurately because a large number of stars will be affected. A similar eclipse in such a rich star field will not happen for another century.'

'I still don't understand why we need to weigh light,' said the man with the flecked shoulders.

'Because either we will cement Newton's place in science, or we shall make the next breakthrough. Either way, Britain will be confirmed as the pre-eminent scientific power.'

The old men exchanged glances. The chairman scratched out a note and passed it to his companions. There was some nodding and shrugging before the chairman fixed Eddington with a disdainful look and announced his decision.

Dyson jumped up from the bench on seeing Eddington. 'Well?'

It took a moment for Eddington to find the right words. He was trying hard to remain composed. Eventually he squeezed out the words. 'One more year's reprieve.'

Dyson grinned widely and gripped Eddington's upper arms. 'You can do the eclipse.'

Eddington smirked. 'We must prepare. We have science to perform.'

17

Berlin
1919

As millions of soldiers and civilians had before it, so the war died. The collapse began at Amiens when the Allied forces opened a gap in the German lines. On 8 August, 1918, thirty thousand Germans were killed, seventeen thousand were captured and the Allied line advanced over ten miles. Panicked, the Germans began to retreat.

Fresh offensives at Albert, Noyon and Arras all yielded similar results. The Germans were in disarray and, as the allies began to breach their last line of defence, an invasion of Germany looked certain. To prevent such a disaster, the Kaiser and his advisers were prepared to agree to almost anything.

They did just that in a railway carriage in the forest of Compiègne on 11 November 1918, in the form of the Armistice. For the German civilians it may have averted invasion but it did not end the hardship. It worsened it.

Elsa surveyed the meagre groceries laid out on the kitchen worktops. 'I thought the end of the war would mean the end of our troubles. But, look, this is all I could get. The prices have almost doubled since November. If this goes on, I don't know how we'll eat. I won't be able to buy enough.'

Einstein looked towards the window. 'I don't know how long the government can last.'

There had already been one revolution, back in the winter at the turn of the year. The rifle shots had set the apartment windows rattling.

'It's not the government I'm worried about.'

'I know,' said Einstein distractedly. Food parcels from his friends were more likely to go missing these days than arrive. He had run out of tea in his own kitchen and shuffled over the landing to Elsa's apartment. He had been about to return with the steaming cup when she returned from shopping. 'I'm doing all I can. I've instructed Mileva to move back to Germany.'

'What?' Elsa's black eyebrows lifted.

'Don't be alarmed. There is nothing for you to be upset about. This is a matter of practicality. I've warned her that I will not be able to keep up the payments to her if the mark keeps falling. Soon the bank notes will be worthless.'

If only the Nobel would come his way, then the prize money would pacify Mileva once and for all. Yet again he had been nominated and passed over by the committee in Stockholm. Why were people so loth to accept relativity?

It was enough to trigger his stomach complaint again. The ulcer was little more than a memory now, but occasionally he thought he could feel it returning.

Elsa stepped towards him. 'Let us marry, straight away.'

'You know I can't do that. The divorce says two years.'

The terms had been agreed and the paperwork completed earlier in the year.

'We could do it quietly. She need never know.'

'Two years, Elsa. 1921 will be our year.' He looked at her beseechingly. 'Why be in such a hurry?'

'By 1921 it will be nearly ten years since you first told me you loved me.'

Einstein felt his cheeks redden. Was it really that long? He lifted his drink to his lips, but it was too hot.

'Besides,' Elsa continued, 'I think Ilse and Georg are getting serious.'

'Georg?'

'Georg Nicolai?' She deployed the mocking tone whenever he was having a lapse of memory and she wasn't sure whether his absentmindedness was real or feigned. 'The doctor, you must remember him. You wrote the peace manifesto together at the beginning of the war.'

'I know. But really? She hasn't said anything to me.'

'Why else do you think he's been visiting so much?'

Einstein shrugged; the question had never really entered his head. 'I'm sorry Elsa, I have other things on my mind.' He blew ripples across the surface of his tea.

'As usual.'

'Why don't we hear something? Why don't they write?' he complained.

Eddington and the English had sent out two eclipse expeditions in 1918, one to Brazil and the other to Africa. That was all that Einstein knew; since then, nothing. He didn't even know whether they had seen the eclipse. Both sites could have been cloudy. All might now be forgotten, but no one had thought to tell him.

Perhaps he would not have been so on edge if the Americans had been successful. They had sent a party to the eclipse under someone called Curtis who claimed to have measured no deflection at all. But he had averaged all his results together. One photographic plate had been in almost perfect agreement with relativity, yet all people were hearing about was the failure.

Planck had arrived in his office to offer commiserations.

It could have been anything, a speck of dust or a flaw on the plate, an incompetent observer or a clumsy analysis, thought Einstein, smarting. But no, if there is any doubt to be had, it's always laid at my door.

He finished his drink and stalked back to his study, where he threw himself into his chair. He felt the sudden urge to cry: ridiculous in a grown man. He had not cried since the day Mileva and the boys had left. That too was ridiculous. They were divorced now. That should be the end of it, yet she seemed to occupy his thoughts more than ever.

He heard someone on the tram the other day laugh like Mileva used to, and it had upset him for the rest of the day.

That evening he watched Ilse laying the table for dinner. The delicate curve of her wrists matched the fragile proportions of the crockery.

'What hopes do you have for marriage?' he asked.

She appeared a little embarrassed. 'Happiness and companionship.' Then with a sideways look to the kitchen, where her mother was preparing the meal, she added, 'Love and excitement.'

Her eyes glittered and Einstein felt quite breathless.

'Herr Einstein, I saw you as I got off the tram. I've been following you.' The man stood in the university courtyard. He wore an immaculate three-piece suit, its pale grey complementing his glowing complexion and short black hair. His eyes were close-set and burned with some deep passion.

'And who might you be?' retorted Einstein. He felt at a distinct disadvantage in his shabby cardigan; when he looked down he saw that he had buttoned it out of sequence.

'My name is Kurt Blumenfeld.'

'Are you an astronomer?'

His eyes creased, amused by the thought. 'No, but you and I have a common interest at heart. We are one tribe, as I think you would say.'

'I see clearly that we share Jewish roots.'

Blumenfeld nodded. 'Perhaps there is somewhere less public that we could talk.'

Blumenfeld stirred the milk into his coffee with a single confident stroke. He set the spoon down but made no move to lift the cup. 'I represent the German Federation of Zionists.'

Einstein froze. 'And why have you come to see me?'

Blumenfeld smiled. He took a leisurely sip of coffee. 'I would have thought for a man like you, that would have been obvious.' There was an aura of danger about him, and opportunity.

'I know that the Federation wants a home state in Palestine,' said Einstein, 'but I'm opposed to all forms of nationalism. I see no fit between us.'

'Herr Professor, your behaviour is more fitting to a politician than an academic. I believe that you are Jewish to your core. I think you are a Jew before you are a scientist, even before you are a man. I can see it runs deeply in you.'

The perception so unnerved Einstein that he put down his cup. 'I'm not indifferent to my fellow tribesmen. But I wonder why you choose to single us out in this way? Now?'

'If once we scattered from Zion to stay safe, now we must return. Hatred is everywhere. If we're to survive, we need a homeland, a fortress. Your position as an academic insulates you to some extent but not everyone is so lucky. Your fellow Jews are no longer being served in all the restaurants and cafés in Berlin. Employment for them is dwindling; they are the last to be recruited and the first to be laid off. A crisis is coming.'

Einstein knew it was true. Ever since the Armistice, Berlin had been changing. Troops had returned home in what remained of their battalions, and had paraded victoriously through the streets to cheering crowds, while in the backstreets Jews had been spat on and called cowards. Beatings and robberies were increasing, and were largely being ignored by the non-Jewish community.

Germany was convinced that it had been wronged, not just by the surrounding nations of Europe but from within – a creeping rot within. And the rot had a name: Jewry.

'You know I'm right,' said Blumenfeld. 'We're unique in the world. We're a culture, and a religion. We can choose one without the other, or both, and still we remain Jews.'

'But why shackle us with the running of a country when we can be free to develop wherever we may? Let us leave the hard graft of agriculture and diplomacy – and the pitfalls of nationalism – to others. I see no reason why a man cannot retain his culture but live in another place.'

'Herr Professor, you can refuse this fight if you want, but that doesn't mean the fight will refuse you. Anti-Semitism is not a rational thing.'

'As much as I have sympathy for your ideas of kinship, I don't feel I can share all of your goals. They remind me too much of jingoism.'

'Jingoism?' Blumenfeld looked temporarily hurt. 'We're not all of the same ideology. Within the movement there's a variety of feelings and beliefs. Yet we all share a single goal: the creation of a homeland state. A place where we can be free. Maybe even a place where we can build a new university.'

Einstein searched in vain for any flicker of hoax. 'A new university?'

'Why not? We will be free to live, to love, to learn. You can help us achieve that freedom, Herr Professor.' The man's face was a study in sincerity.

'Call me Albert, please.' It was a delaying tactic. Despite everything Einstein thought he believed, the man's words were taking root inside him.

'Albert . . . you feel it in you that the Jewish people are distinct and should have a place in the world, don't you?'

Einstein closed his eyes to shut out the intense brown ones looking at him. He then nodded deliberately. 'Perhaps one can be an internationalist without being indifferent to members of one's tribe.'

'So you'll join us?'

Einstein opened his eyes to see Blumenfeld's face taut with anticipation – the first indication of doubt he had seen there. He kept the man waiting a moment longer. 'I'm no Zionist, but I will help you.'

A week later, in an open-topped automobile, Einstein and Blumenfeld drew up to a set of tall metal gates. The handbrake rasped as Blumenfeld applied it and they waited for the liveried gatekeeper to unlock the entrance.

'I've never much cared to drive,' said Einstein, over the idling engine. 'Strikes me that it requires thought that could be devoted to other things.'

Blumenfeld rubbed his hand across the steering-wheel. 'I borrowed it especially. I though Herr Rathenau of all people would appreciate it, rather than have us turn up on foot.'

The gatekeeper pulled open the gate, Blumenfeld released the brake and the car jerked into motion.

Walther Rathenau's house was a mansion, hidden behind trees and high walls. A sweeping staircase led to a vast doorway where the man himself stood, face lifted to the evening sunlight. He seemed taller than Einstein remembered from the Habers' dinner at the beginning of the war, but the swagger was the same. His shoulders were wide and his greying beard was trimmed into a sharp point.

Doubt caressed Einstein's spine.

'Thank you for setting this up, Albert. Do you think we will persuade him?' Blumenfeld whispered as the engine coughed to a stop.

'No, but that doesn't mean we shouldn't try.'

It was late. Einstein and Blumenfeld were clearly not going to be leaving until the early hours. Following dinner, they had adjourned with heavy stomachs to sit around the fireplace, its empty maw hidden behind a decorative brass guard.

'Is anyone cold? Shall I have the fire lit?' asked their host.

Of all of them, only Rathenau could afford to burn wood in the summertime.

The leather armchairs they were seated in were deep and comfortable, and Einstein could imagine the heightened aroma a fire would bring out. He thought of it blending with the tang of his pipe tobacco.

Rathenau sat with ostentatious casualness, cradling a bulbous glass of brandy in one hand and flicking his cigarette ash with the other.

Blumenfeld was in no such condition. The conversation had ranged widely but Rathenau had been stalling, and Blumenfeld was growing impatient. Einstein knew it was time to bring the matter to a head, otherwise they would still be there at dawn.

'Walther, Kurt here does have serious business to put to you.'

'I know, I know.' Rathenau took a sip of his brandy. 'But I'm not about to be recruited to the cause. I cannot exchange my instinctive national loyalty for a newly invented one.'

Blumenfeld cocked his head. 'With respect, Herr Rathenau, you're a Jew. You cannot deny it. You're a Jew and a capitalist.'

'A capitalist. You make it sound like a term of abuse.'

'Jew. Capitalist. It is to the German people.'

'What makes you think you can speak for the German people, I wonder? I'm the one who kept this country in the war. Without my efforts, Germany would have been over-whelmed years earlier. Foreign soldiers would have overrun our lands. Only the longevity of the war spared us. Towards the end they were as sick of fighting as we were. It made the surrender easier by preventing them from ripping us to shreds. Allowed us to keep our pride. And now I'm to be the minister for reconstruction. I'm to be in the cabinet. We should be celebrating, don't you think?'

Einstein thought he must have misunderstood, but the alarm on Blumenfeld's face confirmed that he had heard correctly.

'A member of the government?' asked Einstein.

'Disaster,' muttered Blumenfeld.

'Disaster?' Rathenau's voiced raised a notch. 'A German of Jewish descent will now rebuild this country and you call it a disaster. It's a triumph. This is a new beginning.'

Blumenfeld was shaking his head. 'You will be accused. All Jews will suffer for this. They will twist it to make it look like a takeover, another betrayal. First Jews sabotage the war, and then they make a grab for power.'

'What I do, I do for Germany. Everyone who knows me knows that.'

'Most people don't know you. They see your picture in the paper and they see a Jew.'

'I don't deny my heritage, but love for my Fatherland supersedes the flesh and blood of ancestry. Germany is in my very spirit.' He took a long pull on his cigarette.

'You will put yourself in the firing line, Walther,' said Einstein. 'The anti-Semites will surely target you.'

'Let them. My record with this country speaks for itself.'

Einstein made a dismissive sound. 'They will not be open to reason.'

'A Jewish state could put pressure on countries everywhere to stamp out anti-Semitism,' cut in Blumenfeld. 'We could be a force for good.'

Rathenau pursed his lips as if considering the suggestion but ended up shaking his head. 'I can offer you kinship, but I cannot join your cause.' He took a deep swig of his brandy. 'Nor, I'm afraid, can I wish you good luck.'

Haber's eyes, slightly bulging at the best of times, looked fit to burst from their sockets. 'What are you doing, Albert?'

Einstein slowly put away his pen. 'I'm writing a letter, Fritz.'

His deliberate calm did not take the heat out of Haber's assault.

'Fraternising with those men. They'll undo us all.'

'Do you mean Kurt Blumenfeld?'

'Rathenau is furious. How could you take that man to meet him? What were you thinking? The Zionists will set us back years – decades! – going around with their separatist notions.'

'He didn't appear furious the other night.'

'Well, he was, believe me. Honestly, how could you even think to have asked him? He's our figurehead for acceptance, living proof that we can all exist together as Germans, regardless of race.'

Behind him, Ilse appeared at the study door and reached for the handle. As she did so, she flashed Einstein a sympathetic smile and closed the door.

'Sit down, Fritz.' Einstein gestured to a chair with a pile of papers on the seat.

'I prefer to stand,' said Haber flatly.

Einstein stood up as well.

'Albert, you have embarrassed me. I'd given your name to the German Citizens of Jewish Faith. They were planning to invite you to speak to them. I thought the least you could do was show your allegiance with your fellow Germans.'

Einstein hooked his thumbs in his cardigan pockets. 'I'm not a German.'

'You were born in Ulm. You live in Berlin.'

'My passport says Swiss.'

'You can't erase your nationality by changing your passport.'

'And you can't erase your Jewishness by going to church on Sunday.'

Haber set his chin at the accusation but stayed silent.

'Yes, I've heard you converted. How can you think that it will help in any way?' Einstein's temper grew with every passing word. 'If anything, it makes you laughable to them. Everyone can see what you are; your features give it away. You're nothing but a pussyfooter, neither one thing nor the other. How can any Aryan respect that? And how can any Jew respect you now?'

Haber's eyes were as cold as ice. 'I may be many things, Albert, but I'm not a traitor to my country.' With that, he turned and left.

Einstein was still brooding when Ilse crept into the room. She silently slid a cup of tea along the leather of his desk. She was about to creep out again when Einstein spoke. 'Ilse?'

She looked round expectantly.

'Thank you. You cheer me up.'

She smiled and he envied Georg Nicolai. She must smile like that at him all the time.

Einstein and Ilse strolled home as the afternoon gave way seamlessly to a summer's evening, with no darkness and no drop in the temperature to herald the change. As they walked, their arms were almost touching.

'How are things with Georg?' he began tentatively.

She seemed to frown. 'Good, but he's such a tease. Sometimes I'm not sure where I stand with him.'

'How so?'

'He torments me. Told me the other day that if you had any sense, you would marry me rather than Mama.'

The breath caught in Einstein's windpipe. 'So, he has seen it too.'

Ilse's frown deepened. 'Albert?'

Einstein stopped and faced her. 'I've felt something for you for a while, beyond familial love, I mean. I've been trying to pretend it doesn't exist, but if others can see it too then perhaps I must admit to it. You're a beautiful young woman, so strong and . . .'

'Albert, I am but a girl. Do not burden me with this.' She stepped round him and hurried on.

Upon arrival at the apartment block, Einstein would usually have retreated to his study for a final hour of work before supper, but tonight he hung around in Elsa's living room, where Ilse had settled in the window seat with a book.

Their conversation was unfinished and nagged at him.

In the kitchen, Elsa was busying herself with the final supper, preparations and Margot was laying the table. He waited for her to finish and return to her mother.

'I meant what I said, Ilse.'

She looked up from her book.

Einstein talked before his courage failed. 'I'm not so old that I can't remember what it is to be young. My body is forty but my spirit is strong.' He moved to her and touched her soft cheeks with the dry skin of his fingers. 'No man will ever love you as much as I do.'

'But you're to marry Mama.'

'She will understand. She wants only the best for you and me. I think that I would like another child. Your mother cannot give me that.'

She gasped and removed his hand from her cheek. 'I think perhaps you're right when you say no man will love me like you do . . . but I don't want a child with you. I don't feel like *that* towards you.'

'Ilse, I love you.'

The room changed, as if a chill breeze had passed through. Ilse's eyes widened and she pulled her hand from his. Elsa was standing in the doorway, holding a steaming tureen with a tea towel.

'Mama,' said Ilse, pleadingly.

'Dinner is ready.'

It was a silent affair, save for the tense requests to pass the bowls around. Both girls retreated to their bedrooms as soon as possible, leaving Einstein back by the window and contemplating the checked pattern on his slippers. Elsa sat with her chin in her hand, eyes focused somewhere beyond the walls of the apartment.

He said, 'I suppose I've just had a lesson in relativity.'

Elsa turned, bringing her eyes to a stony focus on him. 'How so?'

'I still see myself as a young man, Ilse sees me an old one, and to you I think I look like a stupid fool. Somewhere there will be a mathematical transformation that will allow all three points of view to be equated, and shown to be descriptions of the same thing.'

Unexpectedly she stood up from her seat, perched on the arm of his chair and curled her arm around him. He instinctively leaned into her. She lowered her voice. 'If you are a stupid fool, at least you're going to be my stupid fool. And perhaps I can stop you being quite so stupid again.'

'And how do you propose to do that?'

She planted a kiss on the top of his head. 'You worry about the universe, I'll see to everything else. Now, let me go and talk to Ilse, then we'll hear no more about it.'

'I've got serious misgivings about these new quantum ideas . . .' Einstein tapped a paper that was sitting on his desk and glanced at Planck.

The elder physicist was sitting across from him, looking perplexed. 'But you cemented our belief in the quantum.'

'I just picked up your ideas and found they worked.' Einstein ran a hand across his brow.

Planck was referring to 1905, the year of Einstein's breakthrough paper about special relativity. A second triumph had been Einstein's explanation of why a sheet of metal would spit out electron particles when illuminated by certain colours of light. Only if the coloured rays were made of packets of light energy, quanta, could the strange effect be understood mathematically. The work had captured his colleagues more than relativity, and continued to do so, much to Einstein's dismay. 'What I didn't realise at the time was all the problems quantum theory would create.'

'The atom needs explaining,' Planck said patiently.

'Indeed, but I don't like the way chance seems to be involved.'

'Is it really that hard to think that some physical processes are random?'

'Yes,' said Einstein indignantly. 'A particle can't think for itself, and the quantum laws can't predict the direction in which light is emitted from atoms. The whole thing must be wrong.'

Planck sighed heavily. Before the discussion could continue, the sound of quick footsteps in the corridor drew their attention.

'Problem?' asked Planck.

Einstein peered through the open doorway to where Ilse was working in the ante-office. A sense of premonition gripped him. Only telegram boys moved that fast.

His heart fell into step with the footfall as it grew in volume.

A puffing lad appeared before Ilse. 'Telegram for Herr Einstein.'

Ilse accepted the message and the lad shot off.

Einstein was already on his feet when she brought it through. Planck rose too, guessing. 'The eclipse?'

Einstein's hand trembled as he slipped the telegram from its envelope. He looked from Ilse's wide eyes to Planck's curious ones and back again, then read the typed message twice to be certain. A calm, almost fatalistic air encircled him.

'Well?' urged Planck.

'It's from Lorentz. He's heard from the British.' Einstein swallowed hard. 'They see a deflection.'

Ilse jumped with a squeal of delight.

Planck shook his head in disbelief. 'General relativity proven.'

'Surprised, Max?' teased Einstein.

'Aren't you? No, of course not. You've never had a moment's doubt. Even so, you must admit that it's good to have confirmation.'

The tension that had gripped Einstein when he first heard the telegram boy's approach dissolved into warmth. It spread through his muscles, loosening them. His cheeks began to ache and he realised that he must be grinning like a loon.

Ilse wrapped herself around him from the side, trapping his left arm in a clumsy embrace of triumph. Self-consciously she let go of him again. 'Well done, Albert. Congratulations. I can't wait for you to tell Mama.'

'Tell your Mama? I intend to tell the world. From now on, no one can have any doubt.'

The scraping of a chair from the neighbouring office drew his attention, reminding him of his sole other employee. He steeled himself to go and break the news.

'Excuse me,' he said and headed across towards the ante-office. Freundlich met him at the doorway, his face a mask. After a moment, the astronomer awkwardly proffered his hand.

'Congratulations, Albert, this is a great moment for you.'

Einstein was about to reassure him that there was much work for Freundlich still to do, and greater glory for him as well. He clasped the hand and shook it firmly, but when he opened his mouth to deliver his words of consolation, he paused and thought again. Then he said simply, 'Thank you.'

Einstein burst through the apartment door and in one motion shrugged off his jacket, removed his hat from his head and took the violin he was carrying from its case. It did not turn into the graceful display he had imagined, but, with some frantic arm-waving to extricate himself from the sleeve, he soon had the instrument under his chin.

Elsa appeared, a look of bemusement on her face.

Rather than talk, he began a version of the Wedding March with more enthusiasm than accuracy. He also began to jig.

Elsa chuckled. 'I don't recognise the violin.'

'I just bought it,' he said, continuing to work the bow. 'I'm celebrating. Marry me, Elsa.'

She looked at him as if it were one of his jokes. 'I thought we had already decided.'

'No, I don't mean in 1921. I mean now, right now.' He stopped playing and stood there with his arms outstretched. 'I'll convert your attic into my study and move in with you.'

The force of her embrace almost knocked him from his feet.

Elsa's wedding ring glinted in the sun as she sifted through the newspapers on the dining table.

'This one's from New York.' She passed over a cutting with an impish look.

Einstein took the sheet and his head rocked back at the headline. 'Light all askew in the heavens! Men of science agog!' He thought he would cry with laughter.

Einstein Theory Triumph. Stars not where they seemed or were calculated to be, but nobody need worry. A book for 12 wise men. No more in all the world could comprehend it, said Einstein when his daring publishers accepted it.

'Where do they get this stuff?' Einstein tutted, but he was unable to summon any genuine annoyance. His eyes fixed again on his name in the headline.

'It's in all the papers.' Elsa spread them about. 'Listen, "One of the greatest – perhaps the greatest – of achievements in the history of human thought".'

'Who would have thought there could be this much interest in my little theory?'

There was a knock at the door.

Elsa pushed her hair into place and hurried to the hall. Einstein ambled along behind.

Standing on the doormat was a sandy-haired postman with a freckled forehead. 'The lift isn't working.'

Einstein frowned at his insolence. 'The electricity costs too much these days.'

'I've got some letters for you.' The postman pointed down the stairwell.

'You couldn't carry them up the stairs?'

'The sack's heavy.'

Einstein followed him downstairs, shaking his head and muttering.

In the hallway the postman pointed to a bulging canvas sack.

'Aren't you going to at least hand them to me, or do I have to pick them out myself?'

The postman looked at him wearily. 'These are all yours.'
Einstein stared at the overflowing contents. 'All of them?'
'All of them.'

He heaved the sack upstairs, developing some sympathy for the postman in the process, and buried the newspapers under the deluge of correspondence. They slid across the polished wood of the tabletop and spilled on to the floor.

'How will I ever answer them all?'

He picked up letter after letter. Most were addressed with some variant of *Albert Einstein, Berlin*. 'Let us separate out the ones that are properly addressed. That way, we may deal first with the ones from people who know us.'

Elsa grabbed a handful. 'Some of these are from America!'

'Universities?'

'I don't know.' Elsa shrugged.

'Perhaps I could do a lecture tour, raise some money.'

They sorted the letters into piles, but Einstein couldn't resist ripping open one with a British stamp. 'It's from Eddington.' He scanned the contents, growing breathless with excitement at what he read.

It is the best possible thing that could have happened for scientific relations between England and Germany. I do not anticipate rapid progress towards official reunion, but there is a big advance towards a more reasonable frame of mind among scientific men. Although it seems unfair that Dr Freundlich, who was first in this field, should not have had the satisfaction of accomplishing the experimental test of your theory, one feels that things have turned out very fortunately in giving this object lesson of the solidarity of German and British science even in time of war.

'I must reply to this one at once,' he said, dashing to his study.

18

Louvain, Belgium

Lemaître hurried down the cloisters, vestments flying, dodging the other seminarians and drawing looks. The morning shadow of the clock tower lay across the courtyard and the bell in the apex began tolling the hour, confirming just how late he was.

With a sinking feeling, he spotted Father Luc dead ahead.

'Overslept again, Lemaître?'

'No, Father, I was working and lost track of time.'

'Working?'

'After morning prayers. On these.' Lemaître opened one of his textbooks.

'That looks like some fancy mathematics on those pages,' said the priest, 'I don't recall it from any of your classes here.'

'Don't be angry, Father Luc. I can't help being so interested. It seems such an elegant way of understanding God's realm.'

'So this is why your grades are slipping. What is it?'

'I have been working through a book written by the English astronomer Arthur Eddington. It describes a mathematical theory of gravity called general relativity.'

Father Luc scanned the books in Lemaître's hands. His eyes locked on one title. '*The Physics of Einstein*? The German we've been reading about in the papers? The German!'

'It's not what you think, Father.'

The towering priest made a gruff noise. 'Are you not being taught all you need to know in your classes?'

Lemaître cradled his books and papers. 'I'm taught so much here about the Lord and his works that it inspires me to look further. I believe there are passages in the Bible that presage modern scientific knowledge.'

'And you think a German Jew can fill in the blanks, do you?' Father Luc glared at him. 'A German Jew who, from what I have read about him, doesn't even believe in God. Hand me those papers.'

One look into the slate-grey eyes and Lemaître meekly complied. 'It's not evil, Father.'

'Now, be off to your lessons. And pay good attention today. No more of this daydreaming.' Father Luc shook the papers at Lemaître. 'We'll let Cardinal Mercier be the judge of these.'

That night the seminary felt claustrophobic. Lemaître's head felt stuffy, as if he had a cold, and he slipped away to walk the streets of the city. He headed for the solitude of the bomb-sites, preferring them to the more populated areas. There was a stillness to the burned-out buildings and the piles of rubble that was comforting. The damage was so extensive that it would be years, perhaps decades, before the reconstruction was completed.

After some wandering he found the spot not too far from the seminary that he liked, and climbed the mound of debris. Near the top he sat down and rested back on his hands to scan the inky black sky. Then he reached for his cigarettes and lit one.

He felt pleasantly giddy from the tobacco. The sound of someone approaching made him turn. He scrambled to his feet, slipping on the loose debris.

'Calm yourself, Georges, I'm not here to admonish you. I know you slip out of the seminary at night to come here. I've seen you from my window,' Cardinal Mercier said calmly.

In the darkness the Cardinal looked like a skeleton. He moved with delicate determination, as if he weighed nothing, and lowered himself gingerly on to the broken bricks. 'Ouch. I should have brought a cushion.'

Lemaître sat beside him. 'Sorry, Cardinal, I'm a little more padded than you.' He patted his hips. Although the war had been over for more than a year, he still found a full stomach a novelty and indulged whenever he could. 'We can go back if you would prefer.'

'No, no, I want to talk to you here. You obviously come here for a reason.'

Lemaître tilted his head upwards. 'Well, I look at the stars and I wonder. The light from these stars takes years to cross space. I like to think that some of it will have begun its journey to Earth back when Thomas Aquinas was alive. It comforts me. Makes me feel a connection across time.'

'We both share a scholarly affection for Thomas, I think.'

'You flatter me, Cardinal. My baccalaureate in his philosophy is nothing to your lifelong contribution.'

Mercier ran his eyes across the sky. 'One's studies are never truly finished.'

Faint sounds of the city floated around them on the breeze.

'Father Luc came to see me today,' said Mercier.

Lemaître dropped his head to await the judgement. Instead the old man asked, 'Where do you feel closest to God?'

Lemaître drew in his legs and hugged his knees. 'I suppose I should say in chapel at prayer, but the truth is, when I'm out here looking into eternity. Is that so wrong?'

Mercier made a soothing sound. 'We must each find our own way. There's no single path. Remember Thomas and his *Deus absconditus*?'

'The hidden God.'

'Precisely, the more actively you search, the more hidden God will become to you. You cannot prove him like you can some mathematical equation. But if you accept him, you will find that he lives inside you.'

Lemaître thought of the wire-cutting patrols that had retreated from each other without violence that night in Ypres.

'Remember,' the old cardinal continued, 'we can know something exists even if we cannot know the precise nature of it. That is not a failure. We don't need to know the essence of God to know that He exists. Even in the sciences, we can know water exists, without knowing that it is made of two hydrogen atoms for every oxygen.'

Lemaître searched the gaunt face. 'I'm surprised by your knowledge. I thought . . .'

'You thought my mind was closed to the new thinking. It isn't. I may not have your gift with mathematics but I am just as fascinated by the discovery of atoms, and stars.'

'Cardinal, do you think God became weary of us? Is that why He is hidden from us, because He has abandoned us? Perhaps that is why I still feel the need to look for Him. I still feel He must be visible.'

'You will never find proof for Him, and especially not by looking at external things. All you need is already within you. God placed it inside you, along with your immortal soul. But even then you can't go looking for it, you must just accept it, as you would a gift, and you will find that it warms you on the coldest of nights. You don't need to look for unifications in all things. Things can exist side by side and still have meaning. Galileo once said that the Bible tells us how to go to Heaven, not how Heaven goes.'

'But I see the connections. I don't think there's much difference between a mathematician's infinity and the concept of eternity, for one thing. For another, in the book of Genesis it says that God created light and matter together and then Einstein discovered . . .'

A twig-like finger waved him silent.

'If there is a connection, then it is a coincidence and of no importance. The Bible does not teach us science. The most we can say is that occasionally one of the prophets made a lucky scientific guess.'

'I want to believe you.'

'Taking the Bible to be infallible science led to Galileo's trial.' Mercier shook his head. 'A lamentable episode. We must never again have theology pitted against science.'

'Can one believe in both?'

'So long as one is clear about what each can do and what each requires. Your confusion comes from the fact that you conflate the two. Science requires mathematics and proof; religion requires belief and faith. Each of us must decide on the balance. There will be those who live their whole lives in doubt, looking for proof. That's fine for a scientist trying to find the working of the universe around him, but for a soul in search of salvation it will lead to confusion. Even if you see Him in the starlight, it doesn't mean that you can find Him by measuring that light.'

Something seemed to collapse inside Lemaître. 'When I hear you speak, I think that I can be a priest but I'll never be a theologian.'

'A priest is good enough for any seminarian, and you have other gifts, your mathematics. From what I can gather, Einstein's ideas are beyond most men – even beyond most mathematicians. If you can comprehend them, then perhaps you should think about pursuing them.'

'You mean leave the seminary?'

'No. Finish your studies here, become the priest you want to be. But then seek your fulfilment in the sciences. I don't imagine it will be an easy path for you – so much of science these days is hostile to religion – but perhaps you can restore some balance. But, Georges, know exactly why you are doing it. Science, technology and knowledge are ways of learning about God's creation but they cannot lead us to God. He is forever hidden from us in this world. If we glimpse Him at all, it is in the quiet, surprising places, where you least expect it.'

'Thank you, I will think about it.'

'Then I will bid you goodnight and leave you to your contemplation.' With a swish of his robes, Mercier disappeared into the darkness.

Lemaître looked back at the stars. His eyes had adjusted enough now that he could clearly see the Milky Way, hanging low over the sky. The constellation of Cygnus swam along the starry river, the great swan's neck stretched out and its wings spread wide.

'The Bible tells us how to go to Heaven, not how Heaven goes,' he said aloud. 'I like that.'

19

Berlin

1920

Einstein glanced at Nernst from underneath a large felt hat. 'Remember, not a word to anyone.'

'And you remember, if this goes wrong, it was your idea.'

They joined the crowd outside the Berlin Philharmonic Hall. The only clue that this was not a concert performance was the lack of evening gowns and black ties. Yet neither was it a rabble. Einstein scrutinised the sharp business suits and the polished shoes. Then he said contemptuously, 'Bourgeois.'

Nernst looked at him askance. 'You can't criticise them for that. Don't you have wallpaper in your apartment?'

'Elsa has wallpaper. I still prefer not to wear socks.' He glanced down. Beneath his shapeless coat and trousers, sure enough, his naked ankles flashed as he walked.

A billboard on the side of the building was plastered with a black-and-white poster. *Antirelativity Rally organised by the Study Group of German Scientists for the Preservation of a Pure Science.*

Nernst indicated the poster. 'Who the hell are this lot, anyway? Why haven't we heard of them at the university?'

'They're a front for something. Political, I shouldn't wonder. But if they can hire the Philharmonic they've got backing.'

'I don't like it, Albert. Let's just leave.'

'What, and miss all the fun? We need to know who these people are.'

He strode off, leaving his friend no choice but to trip along after him.

They took their seats and let the hall fill around them.

When the lights dimmed a stringy-looking man in his thirties took the stage. He announced himself as Paul Weyland and quickly shrugged off the small hint of nervousness that had accompanied his entrance.

Einstein leaned towards Nernst. 'So that's what this is about. I know him. He's an activist, a right-wing nationalist. This isn't about science, it's an attack on Jewishness.'

Weyland spoke with utter confidence from the stage. 'Relativity is a big hoax,' he announced before proceeding to denounce not just the science but Einstein himself. 'He engages in a businesslike booming of his theory and his name.'

Einstein gripped the arms of his seat and concentrated on holding himself rigidly in place.

'Relativity undermines the very absolutes that countries are built upon: certainty and self-determinism. How can such Jewish science do otherwise? They wander the globe in search of places they can usurp, people they can exploit. They undermine absolutes everywhere they go, in their attempt to confuse and manipulate us. We are a people who have been wronged, and it must never happen again.'

There was an explosion of applause from the audience. One or two even rose to their feet to raise their fists and cheer.

'We are rationalists here,' shouted Weyland. 'If further proof were needed of the grand sham of Jewish science, let us now hear from the highly distinguished physicist Ernst Gehrcke.'

The physicist shuffled on to the stage clutching a sheaf of notes. He almost dropped them as Weyland clapped him on the back. Adjusting the papers on the lectern, Gehrcke twisted the ends of his grey moustache.

'Distinguished!' scoffed Nernst. 'We barely see him these days.'

Einstein couldn't help but let out a cackling laugh as the faltering professor read from his notes. Every time Gehrcke used words like 'absurd' or appealed to the audience's common sense, Einstein would let rip, attracting hostile looks from those close enough to hear him.

Soon, there was a ripple in the crowd and a murmuring amongst nearby audience members. At first Einstein thought it was because of the speaker's obvious failure to hold their attention, but then he noticed the heads were turned towards him. People were nudging each other and passing comments with their neighbours. Gehrcke noticed it from the stage but doggedly continued with his presentation.

'We've been rumbled,' hissed Nernst.

Einstein sank into his seat, trying to hide behind his wide-brimmed felt hat and casting his eyes round the auditorium. People were definitely staring and pointing in his direction.

Nernst was turning his head frantically in search of an escape route. His disquiet made up Einstein's mind. Calmly, he stood and removed his hat.

There was the unmistakeable chorus of his name, harmonised by a flurry of movement. On stage Gehrcke was shielding his eyes from the lights and looking blindly into the audience. Weyland reappeared, poised on the steps leading up to the stage. He fixed his gaze on Einstein, flapped his jacket open and stood with his arms akimbo.

Einstein waited for the audience to hush. 'I have found this evening's entertainment very amusing,' he declared. 'Thank you. You can debate whether you believe in my theories until your final breath, but what you cannot do is make them any less true. It doesn't matter what you think,

Nature will do as the Lord has designed. And not even Germany can change that.'

He looked down at Nernst, who was still wincing at that last comment. 'Come along, Walther, we're leaving.'

In silence, he made his way to the end of the row, forcing those in front of him to stand and allow him past. When one set of knees stayed rigidly put, he fixed their owner with a glare. Eventually the young man stood and allowed Einstein to pass.

He climbed the steps to the exit, Nernst close behind. As the theatre doors swung closed behind them, they could hear the beginnings of a commotion.

'I don't know, Kurt. You ask a lot.' Einstein could see Blumenfeld out of the corner of his eye, watching him like a cat mesmerising a mouse. 'I've already agreed to give a paper in Brussels.'

'Give it another time,' suggested Blumenfeld.

'It's not that easy.' Einstein turned to face him. 'It was for the Solvay conference.'

Blumenfeld looked blank.

'The Belgian industrialist Ernest Solvay?'

Still nothing.

'It's the first time the meeting has been called since the end of the war. It's private, the most prestigious physics meeting in the world. Plus, I have other lecture commitments across Europe.'

Blumenfeld fractionally readjusted the angle of his side plate. 'But you do want to lecture in America, do you not?'

Einstein wondered where that intelligence could have come from. He had in fact tried to put together an American tour, but it had collapsed because he had asked for too much money. Elsa had been furious with him when the refusals arrived and told him in no uncertain terms that he had been greedy. 'You must ask for just a little more than you are worth, not enough to solve our finances in one fell swoop. What world do you live in?'

His hasty riposte that she should take over the finances had led to her spending an hour on the telephone with his address book; the result of which was a lucrative string of European lectures that he was now being asked to give up to go with the Zionists to America.

Blumenfeld tapped the telegram he had brought to Einstein's apartment. 'This is your chance. Dr Weizmann wishes you to accompany him to America. It could be a once-in-a-lifetime opportunity for you.'

Einstein grimaced. 'I've never even met Dr Weizmann. I'm uncomfortable with using my . . . celebrity . . . to pull a crowd. Besides, I'm no orator. Not on the subject of a home state in Palestine.'

'But you must admit that a Jewish university is attractive to you.'

Einstein nodded cautiously.

Blumenfeld leaned forward. 'If you take your conversion to Zionism seriously, then you must do this.'

'Who says I'm a Zionist?' Indecision fretted inside Einstein. A warning look at Elsa prevented her offering any more tea. Blumenfeld had eaten quite enough of her fruitcake as it was.

'You have committed to help us. That makes you a Zionist.' The visitor again touched his fingertips to the telegram on the tablecloth. 'Dr Weizmann is the leader of our cause. He has every right to ask you to do this. And you have a duty to accept. Just being there will help. Lecture about whatever you want. They'll all ask you about your science anyway.'

America. The name was like a siren song. Einstein thought of the places he had heard about – Princeton, Harvard – all of them full of bright minds eager for relativity. He rested his chin in his upturned palms. 'What you say *is* totally logical. I am, somehow, a part of this.' He slowly lifted his head from his hands. 'And, therefore, I see too that it is my duty to accept this invitation.'

The façade of confidence fell from Blumenfeld's face. 'Really?'

'Yes.' Einstein was equally surprised by his guest's reaction.

Blumenfeld gathered his things with some haste. 'I must cable Dr Weizmann at once with the good news. Frau Einstein, perfect refreshments as usual.'

Einstein was contemplating the appliqué on the tablecloth when Elsa returned from showing the excited young man out. 'Well, that's that settled then,' she said gleefully. 'Leave everything to me.'

'Aren't you worried about rescheduling the European lectures?'

'Why should I be? By the time you come back from America we'll be able to charge them even more.' She began clearing away the crockery, humming to herself.

Einstein dropped his chin into his hands again, lifting his elbows only when Elsa whipped away the tablecloth. How was he going to break the news to Haber?

The chemist thrust his fists in his lab-coat pockets, straining the material, and glared at Einstein. He said nothing. He did not need to. Anger was written all over his face. The longer he stood there the more uncomfortable Einstein became. He looked from Haber to the flickering Bunsen burner on the desk, to the flask of bubbling liquid above it, to the embarrassed assistant, who was now slinking from the room.

'I can't believe that you would betray us like this.'

'Betray you?'

'Solvay. You're the only German scientist who's been invited. You could've helped us return German science to the world community.'

'Is that not what I'm doing by going to America?' Einstein knew as soon as the words were out of his mouth that flippancy was not a good option.

Haber's mouth dropped open. 'Do you pretend to be this stupid? You're visiting a country that fought against us in the war to lobby for a Jewish state.'

'I'm most concerned about the establishment of a Jewish university in that state. Why shouldn't our brightest young tribesmen be protected from the hideous prejudices of today?'

'You will destroy everything Rathenau and I have fought for. People in this country will see you as the evidence that Jews cannot trusted – running off to beg for help from German enemies.'

'The war is over. Germany lost.'

'Germany is not over. You can feel the anger in the streets. The more the allies squeeze us with their reparations, the more it simmers. Germany will become stronger because of this.'

Einstein bit his lip. He had to admit that the greed with which the allies had pursued Germany verged on persecution. He wanted Germany punished, without doubt, but there was no excuse for the constant humiliation and gloating. The reparations alone were said to run to almost 270 billion marks, a sum that was far beyond Germany's ability to pay. The central bank was printing more money, inflation was out of control, and the country was essentially bankrupt. Einstein had even heard it said that the debt almost equalled the value of all the gold ever mined, throughout all history.

The snubbing of all other German scientists at Solvay was repugnant. Could he really have so much resting on his shoulders? Was he now seen as Germany's ambassador to science? Profound fatigue gripped him. 'Europe is in more trouble than any one man can fix. From now on, my first allegiance will always be to my own people.'

A look of pure hatred came into Haber's eyes. 'You don't even practise the Jewish faith any more.'

'I feel the culture in my blood.'

Haber turned his back and raised his head. Einstein waited, expecting something more. Perhaps a parting shot?

Nothing came. Haber just waited. Clearly, the next time he turned around he would want Einstein to have disappeared.

The physicist did not disappoint him.

20

Cambridge

For a city Cambridge seemed unnaturally quiet, thought Lemaître as he dodged one of the bicycles that glided through the streets. It was not that he minded the quiet. Being fresh from the seminary, it should have made him feel at home, but he was unsettled by the guardedness of the people, the way they walked around with downcast eyes and hunched shoulders.

He had been told that the station fell short of the city itself, and that when a move had been made to extend the line the university itself had blocked the initiative, determined to insulate the colleges. Whether confounding visitors was also part of the plan Lemaître could not decide.

The day had started with bright autumn sunshine and the promise that summer was not quite over, yet the chill in the air now was warning just as clearly that winter was coming. The evening star was beckoning him on, already bright against the fading sky, and by the time he reached the market square the cold stone of the buildings was transforming into towering shadows. A few stallholders were packing away and Lemaître approached one who was hefting crates of unsold apples.

'What can I get for you?' asked the grubby-looking vendor, tilting the fruit for inspection.

'I'm not buying today, thank you. I'm looking for St Edmund's House.' Lemaître was suddenly aware of how foreign his accent must sound.

The man looked him up and down, then nodded over Lemaître's shoulder. 'You've got to go right through the town and carry on walking out the other side.'

Lemaître thanked him and turned to leave.

'They like to keep your lot at arm's length.' The vendor cleared his throat noisily.

Lemaître paused; it took a moment to absorb what he had just heard: your lot. He walked on without comment.

St Edmund's had come into use in the late nineteenth century, when Catholics had finally been allowed back into Cambridge. The hostel also lodged Jews and various nonconformists. Until then they had all been banned by the Test Act of 1673, back in the time of Newton.

He would have hated me for what I am, mused Lemaître. Newton's bigotry towards Catholics was well known. Shortly after completing *Principia*, his great work about gravity, he had openly defied James II, the Catholic king, and been publicly rewarded with a seat in parliament for it.

Times change, thought Lemaître, *and even if he'd not liked me, I suspect I would have liked him.*

He was thinking about Newton again the next morning as he stood in front of the tall gatehouse at Trinity College, gazing up at its heraldic shields carved into the stone. He looked to the right, to where the rooms ran along the college wing, and fixed on the first window: Newton's old room, the one in which he had been studying when Halley came to see him to ask why Kepler's Laws were true.

With a smile, Lemaître entered the college.

Arthur Eddington was already standing when Lemaître appeared at his office door. He was dressed in flannels, with a tightly knotted tie, and moved so carefully that Lemaître thought his jacket must be too tight.

'How do you do?' Eddington's handshake was surprisingly strong.

Lemaître assumed this was a greeting and repeated the phrase.

His new professor smirked, making Lemaître think he had made a mistake. 'My English is not so good.'

'On the contrary, it's excellent.'

This confused Lemaître even more. Eddington kept looking at him in that peculiar way. He wondered if there was a mark on his jacket, but he could see nothing after a surreptitious appraisal of the black material. Perhaps it was the oil he had used on his hair? There was not a trace on Eddington's; his hair was kept in place by a rather severe cut.

'You wear a Roman collar.'

The priest automatically touched the strip of white at his throat. 'Does that make you uncomfortable?'

'Not at all.' Eddington waved his hand, affecting a casual air a little too much.

'I thought you knew.'

An unfathomable look crossed the older man's face. 'You can sit down if you would like.'

Lemaître looked around him for a chair. There was a plush armchair on the far side of the office. He did not want to put any more distance between himself and Eddington, so he clasped his hands behind his back and tried to relax his shoulders. 'I prefer to stand.'

That look again. 'As you wish.' Eddington sat carefully on the edge of his leather desk chair; the wall behind him was lined with books.

'May I ask something?' said Lemaître to break the deadlock.

'By all means.'

'What was the eclipse like?'

'You've read the report, I take it.'

'Yes. I mean what was it like to be there, to see it, to experience it.'

'No idea, old boy. I was too busy with the equipment to notice.'

'What was the island like?'

'Oh, most of it is covered in cocoa plantations. The rest is forest.'

'The lushness of the place must have been amazing. I've seen photographs of the tropics; the leaves can be big as a man.' Lemaître still found himself marvelling at any greenery after the relentless grey mud of the battlefields. He had glimpsed the watery green of the willows on the Cam that morning and stopped to gaze.

'You fought, didn't you?' asked Eddington.

A coldness gripped Lemaître's heart. So that was it. 'Yes, I did, sir. Does that upset you?'

'Because I'm a declared pacifist, you mean?'

'Yes.'

For the briefest moment, Eddington made eye contact. 'No, that doesn't bother me.'

'But something does. Is it my faith?'

'No, my own faith is as seriously held as yours.'

'Then you have reservations about my scholarship.'

'Far from it.' Eddington's brow wrinkled. 'If anything, I find you a touch intimidating.'

'Intimidating?'

Eddington gave a wan smile. 'The essay you sent me. Most extraordinary scholarship application I've ever seen. I find it difficult to believe that someone can understand relativity as well as you do with no one to teach you.'

'I had your book on the subject.'

Eddington leaned over and patted Lemaître on the shoulder. 'I know, old boy, that's the most confounding part.'

A whole day passed before Lemaître realised that the comment had been another thing he'd been warned about: the British sense of humour.

It was months before Lemaître got the full story of the eclipse trip out of Eddington. It came one afternoon as they were sitting over tea and shortbread in the plush common-room. Their chairs were beside one of the panelled bay windows, and the quiet professor started talking as if Lemaître had just asked his question.

'We began scouting for a site as soon as we landed. From the ship we had seen that the mountains in the middle were constantly misty, but the consulate told us the best places would be to the north or west of the island. The sun's so high in the tropics we didn't need to face south, so we followed the advice. That's when we found Roça Sundy, Senhor Carneiro's plantation – fabulous host, very generous with his wine cellar . . .' Eddington glanced over his spectacles to check if his humour had registered. 'Couldn't have made it easier. There was a walled enclosure on the side of the house, perfect for us. Protection against the wind, you see? The house was up on the hill and the land dropped steeply away to the sea, giving us a perfect view of the eclipse position. So we set up the equipment there.'

'Go on,' urged Lemaître.

'Now, of course, we weren't at all confident, because May was the end of the rainy season, but at the beginning of the second week the season turned and the dry wind set in. It was a blessing and a curse. Soon became a problem because as the ground dried it became dusty. Got everywhere, in all the clocks and over the lenses. We were forever cleaning. We tried putting up covers but it got into absolutely everything. Then, the morning of the eclipse: disaster. There was a tremendous downpour: thunder and lightning all over the island. We couldn't work at all. The rain didn't stop until noon and the clouds didn't begin to thin until about half past one.'

Lemaître knew that the crucial part of the eclipse had been expected just after two o'clock, when the moon should have completely covered the sun and made the stars come out.

'Well, as you can imagine, the rain had turned the dust into mud. We went slithering about, uncovering the 'scope, racing to get everything ready in double-quick time. That's when Cottingham nearly fell down the clock pit.'

'Clock pit?'

'We'd dug a hole in the ground so that the clock weight could drop further than the frame allowed. That gave us thirty-six minutes between needing to wind, which allowed us to set the telescope tracking automatically during the eclipse.'

'And was it clear? It must have been. You announced results.'

Eddington squirmed. 'Not exactly clear. We worked through the clouds. We could see the sun, but we didn't know whether we were going to get any stars at all on the plates. A few minutes after totality ended the sun was in a perfectly clear sky, but that's just how luck would have it. We had to work at night to develop the plates, when the temperature had dropped to less than hellish and we could cool the washing water sufficiently. There was no ice, of course; we had to use earthenware pots and hide them away in the cellars. To my very great relief I could see stars in the final plates. We took sixteen in all, but only the final six showed stars. I'd planned to measure the star positions on the island and cable the results, but we heard that the shipping company was threatened with a strike, so we upped sticks and went to the port pronto. Then, of course, we had to wait until we were back home,' he indicated the room, 'before I could do any more.'

'How did that feel? You'd been away for months. Not knowing if you'd succeeded or not.'

Eddington looked sheepish. 'One night I slipped the final plate from its casing and just looked at it, tried to will the answer out of it. Came to my senses when I thought what would happen if I dropped it or damaged it. So, I packed it away and was patient. When I got back, I set about the measuring and the analysis, and announced the results at the Royal Society. They made me stand in front of a portrait of Newton as I gave the verdict.'

Lemaître got that joke immediately. 'Better to have him behind you, than look him in the eye and tell the world that he'd been superseded.'

Eddington's face softened, the first time Lemaître had seen it close to mirth. 'Quite.'

'Did it bother you that the Americans had claimed to see no deflection?'

'No,' Eddington said sharply. 'Curtis didn't discount the obviously erroneous plates. He just averaged all the results together. That's a poor way of reducing uncertainty. If I'd done that with all of our plates from Principe, and from Brazil, where we'd sent a second expedition, I'd have got the same: nothing. But I made sure I understood what was happening on every single plate. If I couldn't correct for the errors, I discarded the data. You don't think that wrong of me, do you?' For a moment, Eddington looked genuinely concerned.

'Not at all, I've read the paper. I agree with what you did, though I marvel at your patience.'

'What else can you do when you feel something is so right?'

'But what made you believe so wholeheartedly that relativity is right?'

'Because I believe that truth is beautiful. Relativity is beautiful – the mathematics, I mean.'

Lemaître pursed his lips for a moment. 'I'm not sure I fully agree with that. I can't help thinking that Newton's gravitation was more elegant: just a single equation, totally symmetrical.'

'But the concept of relativity,' urged Eddington.

'Oh, I agree there. I no longer think of the universe as an empty void dotted with isolated stars and planets. Now I see a continuous landscape of valleys and contours, in which the celestial objects nestle like villages.'

'A bit fancy for me, but nevertheless I'm intrigued that the whole universe can be captured in a single line of mathematics. But who do we believe is closer to the truth: Einstein or de Sitter? A universe with a mysterious anti-gravity energy in it, or a universe that doesn't seem to want to hold matter?'

'There is a way of reconciling the two,' said Lemaître, suddenly emboldened.

Eddington narrowed his eyes. 'How?'

'What if one could expand and become the other?'

Einstein's equation for the universe was the equivalent of a ball, whereas de Sitter's was a flat sheet. Lemaître had been musing one evening about how the two could be linked. He had thought about unwrapping Einstein's globe, but could find no mathematics to make that work. Then he remembered that he lived on a globe that appeared flat: planet Earth. The ground seemed flat because the curvature of the Earth was so vast. So what if Einstein's model expanded into de Sitter's?

'Expand, you say? Well, there is something that I find peculiar about de Sitter's solution.' Eddington was animated now. 'The equation makes it look static, but if you introduce even a particle of matter, there's a strange effect. It shoots off away from you; any light it emitted would be stretched, turned from blue to red.'

'Is de Sitter's effect real?'

Eddington shrugged. 'Who knows? He postulates an empty universe and we clearly live in a full one.'

'But if this de Sitter effect is real, it means that Einstein's model could expand into de Sitter's, spreading out all the matter until the density dropped to essentially zero.'

'But the universe is eternal. It's always been there, and always will be there. Why would it change?'

'I don't have all the answers, but, mathematically, I think it's possible.'

Eddington lifted an exquisite china cup and took a sip of tea. 'You know, I can't decide whether you're a madman or a bloody genius, but I do know that you can't entertain a mathematical theory without observational evidence. Telescopes are seeing further than we ever imagined. Of course, what holds the astronomers back is that some of them don't understand how to use the equipment properly.'

'You refer to Curtis and the eclipse.'

Eddington gave him a sly grin. 'Too many think that just by looking through the eyepiece they'll see the answer written in the sky. They don't understand the work it takes to transform observations into useable data. But there are others who do know what they are doing: astronomers who are destined to be remembered forever, and if you're going to propose anything with relativity, you need their observations.'

Excitement pulsed in Lemaître's veins. 'Who are they? I've still got money in my travel grant for when I'm finished here. How can I meet them?'

Eddington grinned his lopsided grin. Lemaître had stopped seeing it as a smirk. 'How are your sea legs, old boy?'

21

New York City

1921

Einstein's stomach bubbled, and not from the movement of the ship. He had grown accustomed to that within a few days of leaving port in Holland. No, it was the anticipation of what awaited him that made him queasy today. He eased a hand across his waistcoat, hoping the warmth of his palm would settle his insides without giving away his discomfort.

Elsa had already spent the morning in a fearful fuss, choosing and changing his necktie and collar numerous times, repeatedly checking he had put on socks. Now she was excitedly peering through the porthole, smiling from ear to ear at her first sight of the Statue of Liberty.

'Don't you want to see, Albertle?'

'I will see New York soon enough,' he said, more peevishly than he had intended.

Initially, the isolation of the voyage had relaxed him. Elsa had been housed in her own suite, just across the corridor, which had allowed him to work whenever he wanted. It had been a pleasant way to pass a dozen days, and he had grown increasingly pleased that his ridiculous request to travel steerage had been ignored. At the time it had seemed right to downplay his journey, especially with Haber's objections ringing in his ears. He might not agree with the assimilationists, but he did not want to damage them.

With the ship now edging towards the American dockside, doubts filled his mind about the wisdom of the expedition.

Chaim Weizmann knocked and entered without waiting to be admitted. If Einstein had been able to summon the courage, he would have asked to be alone, but the set of Weizmann's jaw told him it was inadvisable.

The leader of the Zionist movement had something of Charles II about him. His long chin was hidden behind a goatee beard, and a moustache overshadowed the flat line of his mouth. All that was needed to complete the picture was a luxurious periwig to hide the bald pate and the handle-like ears that jutted from his head.

'I've organised with the captain that we can use his cabin for the press conference.'

Einstein looked up sharply. 'Press conference?'

'Of course. There's a score of reporters waiting on the dock. You're going to be in every newspaper tomorrow.'

Einstein fought nausea. 'You know I didn't want any fuss.'

'And you know we need publicity for the cause.'

'I thought a few dinners, functions, lectures at the universities . . .' He raked a hand through his hair.

'Albertle, leave your hair alone,' hissed Elsa.

He perched on the edge of a chair, hands in his lap, trying to untangle his mind. The winged collar felt like a neck brace. 'I can't speak English well enough.'

'You'll have an interpreter. Just remember to emphasise the need for a Jewish university,' said Weizmann. 'No one's going to argue with that.'

'I can't believe I agreed to this. I feel like a prize ox.'

'You're just nervous.' Elsa reached over to flap at his restless hands. 'Stop picking. There'll be photographers. We don't want pictures of you with ragged nails.'

'Quite right.' Weizmann headed for the door. 'I'll come and get you when it's time.'

Weizmann was a scientist, a chemist by training. On the voyage over he had taken an interest in relativity. At first Einstein had assumed it was genuine curiosity, but towards the journey's end he had perceived it as politeness. Weizmann had begun coaching Einstein on what to say about the university.

The physicist suddenly felt too old for all this. Since the incident with Ilse, he had noticed more signs of ageing in the mirror. His face had become dough-like, not that he had ever enjoyed a strong jawline, and his complexion had lost its vigour. His moustache was still dark but his temples were greying. Worse than his physical appearance, however, was the fatigue. He just didn't seem to have enough energy any more.

He turned to his wife; she had made herself up that morning to look glamorous, as if they were going out to dinner. 'I'm not sure I can do this.'

She touched his cheek softly. 'Of course you can. This is the spotlight you deserve.'

He thought for a moment. 'Perhaps you'd better comb my hair.'

The captain's cabin was adorned with antique compasses and barometers that Einstein inspected in an effort to distract himself. There was a small inlaid writing table in one corner, and Einstein moved to sit behind it so that there was a barrier between him and the press.

'Stand in front of it,' said Elsa, 'and lean back. It'll make you look relaxed.'

He dragged himself to his feet and positioned himself in front of the desk. He had just finished rearranging himself when Weizmann arrived with the reporters. Trilby-hatted, notebooks and pencils in hand, they crowded into the tiny cabin, trapping Einstein. He swallowed down panic.

The press were all smiles and toothy grins but, far from putting Einstein at ease, they unnerved him. They were all strangers, yet they were behaving as if they were his lifelong friends. The din of their questions, in a foreign language he could barely understand, was cacophonous.

'Gentlemen, gentlemen, gentlemen!' said Weizmann, calming the pack.

Einstein's heart thumped. He had no idea what Weizmann was saying, so he took his cue to begin speaking when everybody else stopped.

'Despite my most emphatic internationalist beliefs,' he began haltingly, 'I feel an obligation to stand up for my persecuted and morally oppressed tribal companions. The prospect for establishing a Jewish university fills me with particular joy, having seen countless instances of prejudice. People write to me daily of these injustices against themselves, or their sons or husbands. It has to stop. That is why I am here, to secure the support, both material and moral, of American Jewry for the Hebrew University of Jerusalem.'

Weizmann nodded his approval and opened the meeting for questions.

'Please, can you give a one-line description of your theory of relativity?' asked an eager-faced reporter through the translator.

In spite of himself, Einstein chuckled. 'One line! I've been struggling to fit it into an entire book and he wants a single line.'

The reporters smiled, then hushed, clearly thinking it was a joke that would then be followed by the answer.

Einstein felt naked under their gaze. He crossed his legs. 'Very well,' he said, thinking furiously. 'It is a theory explaining the nature of space and time, and it leads to a theory of gravitation. Will that do?'

The next question was directed at Weizmann, who spoke back almost immediately and almost brought the room to its knees with laughter. Einstein looked helplessly at the translator.

'He was asked whether he understood relativity. He said that you explained it to him every day of the voyage, and that now he is thoroughly convinced that *you* understand it.'

Einstein laughed. The next few questions passed without incident, and gradually he relaxed. At least the staccato fashion in which the interview was conducted through the translator gave him time to think.

One question made him look hard at the reporter: 'What do you make of those who attack your theory, especially those in Germany?'

'No one of true knowledge opposes my theory. Those who oppose the theory are animated by political motivations.'

The reporters asked for clarification: 'What political motivations?'

Einstein considered ducking the question but then reconsidered. What was wrong with the truth? He spoke clearly and directly. 'Their attitude is largely the product of anti-Semitism.'

The reporters scribbled furiously.

Weizmann stepped forward, talking English again and apparently drawing the meeting to a close. The reporters looked rather disappointed.

'I hope I have passed my examination,' Einstein said to the translator, who relayed the message.

The reporters laughed once more and Einstein's cheeks lifted into a broad grin. As the chorus continued he found himself wondering if it was really that funny. Nevertheless, he played along until his cheeks ached.

As the last of the reporters were filing out, one paused. He scratched his temple with the base of his pencil and turned back. He said in halting German, 'Mrs Einstein, do *you* understand relativity?'

Everyone fell silent.

Weizmann looked anxious but Einstein calmed him with a glance, then swivelled to Elsa to await her answer. There was the hint of a smile on her face. 'Oh no,' she said with an easy manner. 'Although he has explained it to me many times, it is not necessary to my happiness.'

A flicker of disappointment crossed the reporter's face. Robbed of his headline, he tipped his hat and left without further comment.

Einstein thought he would burst with pride. Elsa looked at him excitedly and he winked back.

Spring had yet to arrive in New York, so Einstein slung on his favourite overcoat. The ill-fitting grey garment felt homely. He picked up his pipe and his violin case. 'I'm not trusting this to the deckhands,' he said, to stifle any possible objection from Elsa.

She was wearing a fur-trimmed coat of excellent cut, tailored to make the most of her matronly figure. She looked almost stately. Einstein was amused by the idea. He sucked on his pipe, wishing that there were some tobacco in it.

Weizmann was waiting on deck. 'That all went well. Everyone was very pleased with you.'

'Much to my surprise, I think I enjoyed it.' Einstein turned to the railings and had to grab his pipe from his mouth, opened in astonishment, before it could crash to the deck. The dockside was a sea of waving handkerchiefs. There were people everywhere, and not just on the flat of the jetty. Men were wrapped around the metal skeletons of the cargo cranes, waving and grinning like monkeys up trees. Car horns honked in answer to the ship's mighty steam whistle.

'My goodness, the Americans certainly know how to welcome a ship.'

'You think this is how they behave for every ship?' Weizmann shook his head. 'This isn't for the ship. This is all for you.'

'I wish I'd made you wear a better coat,' muttered Elsa.

She leaned forward over the rail. Einstein followed her gaze. There were people everywhere – more than Einstein had ever seen gathered in a single place before.

'I can't believe it.'

'You're a genius,' said Weizmann, with a look of satisfaction. 'They might not understand what you've done, but they believe that you *have* done it. And how many times in your life will you get to see a true genius? America likes nothing more than success. Birthright counts for little here.'

'Truly?' The thought was arresting.

Weizmann nodded emphatically. 'Success is all they care about. When they see it, they celebrate it. You're a celebrity.'

Einstein picked out individual faces, amazed at the gawping, the smiles, the jubilation and the astonishment he discerned.

There was a troupe of cheerleaders high-kicking their way through a routine, chanting, 'Einstein, Einstein, Einstein.'

It was then that he realised that something was missing. It took a moment for him to identify it. When he did, the realisation was almost shocking. His stomach was quiet; the leaden weight he had been carrying around inside for so long had evaporated. A wave of emotion swept over him.

Beside him, Elsa was beaming at the throng. Years had fallen away from her, too.

Einstein raised a hand to the air and waved at the crowd. The responding cheer was deafening.

PART III
Curvature

22

Harvard, Cambridge, Massachusetts

Lemaître's first sight of Harlow Shapley was the worn leather soles of his shoes. The director of Harvard College Observatory was face down and immobile on a grassy ridge some way away from a collection of domed buildings.

Lemaître broke into a trot. 'Professor Shapley! Are you all right?'

A stubby finger shot into the air and an intense southern drawl said, 'Don't say a word, whoever you are. Come down here. Not a word.' He pointed to the ground next to him and Lemaître lowered himself, stifling a sneeze as the smell of grass assaulted his nose.

Shapley was holding a pocket watch and peering over the ridge at a column of ants marching to and from an upended tree stump. There was a pair of marks, about eighteen inches apart, scratched in the dirt, and a thermometer set in a clamp beside the tiny creatures. He followed the movement of the insects, checked the watch and scribbled down a figure in a dog-eared notebook. 'I knew it. I goddamn knew it!'

Lemaître laughed at the casual blasphemy. Shapley looked over his shoulder. His large eyes widened at the sight of the clerical collar. 'You must be Georges Lemaître.'

The American vowel sounds still grated. Lemaître had corrected the mispronunciations of the immigration officers and then of the customs officers. Eventually, however, he realised he was fighting a losing battle. He decided to let it go and said, 'How do you do?'

Shapley grinned. '"How do you do?" – wow, you *are* from England.'

Lemaître decided to let that go as well.

The two men shook hands, rather awkwardly as they were both still lying on the grass.

'Eddington says you're quite a mathematician,' said Shapley.

'I enjoy the certainty of algebra. I find it refreshing.'

Shapley returned his attention to the ants. 'Look at this. They march faster in the warmer weather. I've been measuring them for nearly a full year now. I just need another six months to see if the trend repeats and then I can publish.'

The river of insects continued its back-and-forth looting of the dead tree.

'I'm thinking of cutting a hole in the window-frame of my office and building a tank for them, so I can control the temperature.'

'I didn't know you were an entomologist as well.'

'I just do this to relax. Sometimes after a night in the domes, I find it hard to concentrate the next day. So I come out here. Never takes long to find a column. I just pick out one ant at a time and time them crossing the gap. Put them all together and bingo! If only astronomy were that easy.'

The sun was beating down on Lemaître's black clothing. He wiped a bead of sweat from his upper lip.

'Let's get you in the shade.' Shapley reached for the thermometer. 'Can get pretty hot here in the summer. Freezes your ass off in the winter, of course.' He wound an elastic band around his rolled-up notebook and pushed himself up from the ground.

It was only when the professor stood up that Lemaître realised how short he was. The director dusted down his brown tweed suit and looked properly at the new arrival.

'I must say, it's something of a novelty for us to have a physicist here.'

'I'm interested in learning more about astronomy because I think relativity can describe the whole universe, but to test it I need to know what the universe looks like out to its deepest realms.'

Shapley whistled. ' "Deepest realms." I'll be honest with you, Georges – you don't mind me calling you Georges? No? Good. – I don't understand relativity too well. I've always thought that theories were passing things but good observations never fade.'

Lemaître considered his words. 'What about theories supported by good observations?'

Shapley broke into a toothy grin. 'Let me take you to see the computers.'

The main offices were located in a solidly-built whitewashed house, set apart from the huts and domes which, Shapley explained, housed the telescopes and other scientific instruments. Gardeners in wide sunhats tended the hedges that lined the paths between the buildings.

Inside, the dark wooden panelling swallowed the sunlight. Lemaître's eyes adjusted to the gloom as Shapley led him through corridors and offices full of framed images, bookcases and floor-to-ceiling wooden filing cabinets that required ladders to reach the top drawers.

'Here is where it all happens,' said Shapley, opening a door.

In the room beyond there were about a dozen women hunched over desks, peering through magnifying glasses at large photographic plates. Others sat at desks with ledgers, making comparisons.

'Good afternoon, ladies.' Shapley then introduced his guest.

There was a chorus of greeting.

'In here, the stars are the ants.' Shapley pointed to one of the glass plates. It was splattered with dots; each tiny mote was a star. 'Our computers here,' he indicated the women, 'are the engine of what we do. They catalogue the stars, measure the plates and identify the Cepheids for us.'

Lemaître had learned about the Cepheid variable stars from Eddington. They provided the key to measuring distances across space because they varied their brightness in a repetitive way, taking anything from a few days to a few months to complete a cycle.

The way to turn the pulsation into a distance had been invented here by one of the computers, Miss Henrietta Leavitt, and Shapley had used it to define the outline and limits of the Milky Way.

'Is Miss Leavitt here?' asked Lemaître.

There was a shift in the atmosphere.

Shapley explained, 'I'm afraid she passed away a few years ago, before reaching her retirement. We still miss her.'

'May God rest her soul.'

'Why don't I show you the instrument that she used? Miss Cannon, is anyone using the blink comparator?'

A round-faced woman with pearls and a pile of greying hair looked up. 'Not at the moment, Mr Shapley, but Miss Maury is due in there soon. Her plates are already loaded.'

The blink comparator sat in its own small room, on a desk that had been pushed against the window. The blind was down, but two holes had been cut into the slats to allow the summer sun to fall on to two plates suspended either side of a T-shaped piece of tubing that terminated in a telescope eyepiece.

'You put two photographs of the same patch of sky in the machine, one on each side. Adjust them so they line up perfectly. Then the handle allows us to blink between the two views. The variable stars show up because they seem to flicker.' As he spoke, Shapley flapped the tails of his jacket and sat in front of the contraption. He clicked the handle from side to side, squinting into the lens.

'Here, you have a go; there's plenty on the plate. Top left is good.'

Lemaître took his turn. There was a staggering number of stars. They appeared as black spots on the negative plates and, as promised, when he slipped the lever back and forth the variable stars appeared to flash.

'Amazing,' he breathed, 'and this is just one plate?'

'It takes thousands to cover the whole sky.'

'And the computers work like this all day?'

'That's women for you: hard workers, focused, sharp, infinite patience. I couldn't do it. Why the hell do we still exclude women so easily? We've got one girl here who's really special. I've got her on to the Ph.D programme. First in the country I think – maybe the world.' There was obvious pride in Shapley's voice. 'She's English but couldn't get a place there, so she came here. She was a Cambridge woman, Cecilia Payne. Heard of her? No? Shame. Anyway, seems like Eddington's lectures inspired her.'

Lemaître leaned back in his seat. 'Professor Shapley . . .'

'Harlow, please. Let's not be too formal here.'

'As you wish. I'm interested in the spiral nebulae.'

'Not another one convinced they're beyond the Milky Way, are you?'

'I don't know what to think. That's partly why I'm here. I'm intrigued. On the voyage over, I read papers from Slipher in Arizona about the terrific velocities they appear to have.'

'You mean their redshifts.'

Lemaître had read the papers with a sense of growing excitement. Something was turning the light from these enigmatic objects red. The most obvious way to interpret the signals was that they were shooting off into space from the Milky Way and that their light was being stretched and turned red in the process. Yet the speeds Slipher was measuring were almost beyond belief – hundreds of miles *every second*.

Despite his scepticism, Lemaître's mind kept returning to the maths of de Sitter's universe, and the way that when you calculated the distance between any two points they looked as if they were receding from each other. The only trouble was that de Sitter's solution relied on a universe devoid of matter. Could the same be true for the actual universe and real celestial objects?

'I would caution you to take care,' Shapley growled. 'There are a great many things we don't understand about the spiral nebulae. Any conclusion is just a matter of wishful thinking.'

'You debated their nature with Curtis, I believe.'

'I *discussed* it, I presented facts; he speculated.'

'But you don't believe they're beyond our galaxy, do you?'

Shapely crossed his arms. 'I don't believe or disbelieve anything without proof. Let me give you another example. You're a God-fearing man, but me, I don't believe in God . . .'

'I think religious faith is something you either have or do not have. You can't learn it. My only perplexity is with those who are hostile to the very notion.'

'Well then, we have some common ground, because it's not that I believe God doesn't exist, it's that I have yet to see any evidence either way. I need proof. Prove to me that God exists and I'll believe. Until that time, I remain agnostic.'

'Yet you believe that the spiral nebulae are not individual galaxies.'

Shapley snorted. 'It's all down to their distances, and to gauge those we need to find Cepheids in them. But we don't. We don't see any stars in them at all. So I conclude that they're gaseous objects.'

'But the speeds?'

'Prove nothing. And are only speeds if we trust the observations.'

Lemaitre fenced, 'I thought you said observations never fade.'

Shapley raised an eyebrow. '*Good* observations never fade.'

Lemaître spent the summer and autumn getting to know the stars. He learned about celestial coordinates by working in the domes with the other observers. They taught him how to move the tall 'scopes into position, reading from the etched brass setting circles on the mountings. Even though the telescopes were twice the size of a man, the counterbalanced instruments could be pushed with just a single finger to find any star in the sky.

He became adept at loading the photographic plates into cameras to capture the stellar portraits, and he learned how to split starlight into a spectrum and then analyse the dark lines that appeared like fingerprints on the underlying rainbow of colours. Each line was the result of an interaction between the star's atoms and the rays of light it emitted. Together, the pattern could be analysed to tell the composition, movement and temperature of the star. The astronomers seemed to take it all for granted, yet Lemaître would still shake his head in disbelief that such knowledge could be harvested from such distant realms.

By the time the winter constellations lifted themselves above the tree-lined horizon, Lemaître no longer saw the stars as points of light, he saw them as individuals.

He grew accustomed to the out-of-sync lifestyle whereby the observatory staff were sitting down to meat and two veg as the night observers were dipping bread into their boiled eggs.

As he was returning from breakfast one afternoon, a poster on the noticeboard caught his eye. Designed as a vaudeville advertisement, it was emblazoned with the title 'The Observatory Pinafore'.

'Will you be auditioning, Dr Lemaître? It will be so much fun.' The accent was English and belonged to a slim young woman with a mischievous look in her eye.

He answered her with a booming laugh.

'I sang Gilbert and Sullivan back in Cambridge. I'm rather looking forward to giving it another go. It's a spoof to be directed by Professor Shapley himself. His wife's going to be in it, too.'

'Then I wish you the best of luck . . . I hear you attended Professor Eddington's lectures at Cambridge. He was the one who taught me the value of astronomy when constraining physical theories.'

In truth, Lemaître had found another of the dons, Ernest Rutherford, the better lecturer. He had made the subject of splitting the atom come alive, whereas there was little panache to Eddington's delivery beyond the occasional sardonic comment.

Miss Payne's face lit up as she recounted her experience. 'I found Professor Eddington such an inspiring lecturer. His telling of the eclipse trip quite captivated me. Reminded me of my school days. May I tell you a silly story?'

'I'd like nothing more.'

'When I was a schoolgirl, my teacher brought in a map of the world and covered up all the countries. She then taught us about foreign exploration. Whenever she would start on a new country she would uncover it from the map. Her stories were so exciting I used to dream of exploring the places that remained covered. I thought I had discovered what to do with my life. But, by the end of the year, she had uncovered everything. All the countries had been visited. I can still remember the disappointment. Then I started reading about science and fell in love with it. I became almost frantic one day with impatience. I was afraid all the things to be discovered would have been found before I grew up.'

Lemaître smiled. 'I think the universe has an endless supply of surprises for us. It's a shame you had to leave your home country to achieve your dream, though.'

'As have you.'

'Not quite. I have a place waiting for me in Louvain when my travels are finished next summer.'

'Well, I don't think it's such a bad thing for me. This is a country of more possibilities than England, more equality in general. Astronomy in England still seems rather stuffy – apart from Professor Eddington, of course. But here, it's alive. I don't think I ever want to go home.' She grew suddenly embarrassed. 'Listen to me babbling! Are you sure we can't persuade you to audition?'

'Miss Payne,' he said gently, 'I look forward to watching, but the best contribution I can make will be as a receptive member of the audience.'

A month later, the smell of hot, spiced cider wafted around the observatory's lecture hall. An audience gathered, toasting each other and the coming new year. The stage was a simple affair of heavy curtains and a potted plant, but for good measure someone had hung up a framed picture of Saturn. The planet's rings were tilted to their best angle, as if it had been posing for the picture.

Lemaître found a seat next to a stout fellow he didn't recognise. By the turn of the conversation he seemed to be an acquaintance of Shapley's who lived in nearby Boston.

'You are up for the evening specially?' asked Lemaître.

The fellow's ruby-coloured cheeks quivered in mirth. 'I wouldn't miss this grudge match for the world.'

Before Lemaître could ask for clarification, Shapley took to the stage and the audience fell silent.

Dressed in tails, he explained in his precise drawl that the spoof had been found in the observatory's archives, written back in 1879 but never performed. Having reviewed the piece, he was struck by its relevance.

This drew another guffaw from Lemaître's neighbour.

'Some of us have sung operettas before and some of us are new to the game,' continued Shapley. 'What we lack in technique, we make up for with enthusiasm. Let the revels begin.'

The music started, played by a small troupe tucked into the side of the hall, and the computers sidestepped their way on to the stage in time with the music. Dressed in the high-necked blouses and long skirts of their nineteenth-century counterparts, they sang:

> We work from morn 'till night,
> For computing is our duty,
> We're faithful and polite,
> And our record book's a beauty,

The audience responded enthusiastically, laughing along with the music rather than humming. Lemaître himself filled the room with his laugh, particularly when 'catechism' was rhymed with 'prism'.

Miss Payne played one Josephine F. McCormack, a peerless setting-circle reader and target of the villainous Dr Leonard Waldo's attentions. With much fake moustache-twirling, Waldo tried to lure her away from Harvard to work at another observatory. The Harvard staff could not take this lying down, and set about embroiling Waldo in such plots that he failed in his attempt to poach Josephine.

Villain thwarted, Josephine restored, the performers took their applause.

'You've gotta hand it to him, he's a brave man,' said Lemaître's neighbour, elbowing him to draw his attention.

'I'm sorry, you have me at a disadvantage.'

'You don't sound like you're from round here.'

'I'm from Belgium.'

'That explains it. You'll not know of the rivalry between the east and the west coast astronomers, then.'

Lemaître shook his head.

As the audience began to get up, his guest eagerly filled him in. Shapley had previously worked at the Mount Wilson Observatory in California, where the largest telescope in the world was situated. When the war ended, a young astronomer by the name of Edwin Hubble arrived. A rivalry ensued that led Shapley to quit in favour of Harvard, where his promotion to director meant that he could build a department.

'So, you see, Shapley loathes Hubble and vice-versa,' the informant concluded, with a secretive tap on his bulbous nose to indicate the information's confidentiality.

'You think the production was a warning to Dr Hubble?'

'No question about it. No poaching – that's what he's saying.'

Lemaître looked back at the stage, now empty. Somehow it was no longer as much fun as he had thought.

There was a pall over the observatory as if someone had died. It was never a raucous place at the best of times, but today it resembled a mausoleum. The staff tiptoed along the corridors, and Lemaître's greetings were returned with subdued looks and hurried acknowledgements. Even the secretaries appeared to be trying to type softly.

Lemaître saw that Miss Payne's door was open and sidled into her office. 'What has happened?'

She pulled off her glasses and whispered, 'It's Professor Shapley.'

Lemaître's momentary spike of concern subsided as she continued. 'He's heard from Hubble in California. The Mount Wilson team have found Cepheids in M31 and measured them.'

M31: the great spiral whorl in Andromeda. The pained look on her face told him the next bit.

'So, Hubble's done it: proved the spiral nebulae are outside the Milky Way?'

She nodded in the direction of Shapley's office. 'I was in there when he opened the letter. He's not, er, very happy.'

'I can imagine.' Lemaître's mind spun with possibilities. 'It came as a letter, you say?'

'Yes.'

He had to see it; his whole life seemed to depend on it. He looked at Miss Payne's pale eyes. 'Wish me luck.'

Shapley was sitting with his back to the door. His feet were up on the windowsill. 'What is it, Georges? I'm rather busy.'

'I understand you've had a letter, some important results.'

'News travels fast. I'm considering how to respond.'

Lemaître stepped closer; he could see the letter unfolded, lying on the blotter. 'May I read it?'

Shapley waved his stubby fingers over his shoulder. 'Go ahead.'

He read with a growing breathlessness. 'This changes everything.'

The great Andromeda Nebula was so far away that it was far beyond the boundary of the Milky Way, the boundary that had been measured by Shapley himself.

The observatory's director dropped his feet to the floor with a bang and spun the chair round. 'Don't get too excited.

It's the most entertaining piece of fiction I've read in a long time.' His expression was stormy.

'Even if you make small corrections it won't change the outcome,' said Lemaître. 'The distances are so huge. Andromeda is a system of stars in its own right, a whole different . . . galaxy. Another Milky Way.'

'That's what others are saying. Not me.'

'But the evidence . . .'

'The evidence is not conclusive. I'm telling you, Georges, don't trust Hubble.' With that, he swung the chair back and continued to brood. Clearly the conversation was over.

That night, Lemaître left his small room and walked deep into the countryside. Leaving the trees and buildings behind, he shivered in the cold. He waited for the light to drop and his eyes to become accustomed to the dark. One by one, the stars brightened into view.

He found the square of Pegasus and traced the stars round to follow the curve of Andromeda. Stopping at the third star, he moved his gaze upwards to the next of the guide stars. Now he needed to look just above and to the left. He peered at the sky until his eyes began to smart.

And then he saw it. A tiny patch of light hovered on the limits of visibility: there one moment, gone the next, but unmistakeable nonetheless. It was the Andromeda galaxy, the ancient light of its billion stars arriving at Earth after a journey of almost a million years, according to Hubble's calculation.

Lemaître's eyes tracked across the other stars in the night sky. All of them beacons in the Milky Way, our own galaxy. For millennia it had been thought to be the whole universe, now it was clear that it was just part of an archipelago. Andromeda, far beyond, was another island of stars, and there were yet others that the astronomers were finding spread across the night sky, stretching into the very depths of the night, thousands of other galaxies.

He bathed in the starlight, marvelling at the silver landscape around him. In his rapture he forgot about the cold. Instead his whole body tingled with excitement, and in that instant he knew what he had to do.

Next morning he noticed Shapley leaving the main entrance. The director was carrying his rolled-up notebook and thermometer, his short legs moving rapidly. Lemaître hurried after him as the director veered off across the grass. Unable to catch up, he called out, 'I plan to go and see Hubble.'

'Fine.' Shapley did not break his stride.

'You don't mind?'

'How you waste your travel fund is not my concern, Georges.'

Lemaître stopped following. He was about to turn away when Shapley halted. The little man came marching back to him.

'From the moment he arrived at Mount Wilson, he wanted me out. Did everything he could to curry favour with the governors – criticised what I had achieved, bragged about fighting in the war, pretended to be British. He was born in Missouri – same as me – yet because I still have my accent, he treated me as an inferior. He's a social climber. Thinks he can be a celebrity and will stop at nothing to become one. Don't believe a thing he tells you about me. He'll tell you I lost that discussion with Curtis, but I didn't. Think about what you're so ready to believe in, Georges. You want it to be true, so you believe. We both know you have a predisposition to belief.'

'My faith doesn't enter into this. I've learned to keep it entirely separate from my science.'

'We'll see. They'll tell you I lost. But I was playing a cleverer game than they knew. I couldn't stay in California with his constant agitation, so I applied for the directorship here. I knew I was too young, so I had to impress the governors some other way. The debate was the place to show I was a steady pair of hands, capable of shouldering responsibility and not being reckless. I wasn't going to get involved in a debate where there was no clear answer. Debates do not decide science, observations do. So I presented the accepted observations and let Curtis shoot the breeze.'

'But there are new observations now, and I need to go and see them for myself.'

They locked eyes.

'Think of history,' said Shapley. 'Astronomy has made major advances by removing us from the centre of the universe. Copernicus and Kepler proved the Earth is not the centre; I proved our solar system is not the centre of the galaxy. But with these spiral nebulae all blasting away from us . . .' He shook his head. 'If you insist that these things are all moving away from us, you inevitably put us back in the centre of the universe again. It's a retrograde step. Don't make the same mistake, Georges. If you do, your credibility will be ruined.'

'What if I told you I had a way of avoiding that mistake, using relativity?'

Shapley gawped at him. 'You're serious?'

'Yes.' He could not prove it yet, but he was seeing the beginning of it and knew that long calculations lay ahead.

Shapley barked, 'Fine. Go and see Hubble if you must, but remember one thing.' A sly look crossed his round face. 'I may be sceptical of theories, but he positively hates them – and theoreticians.'

23

Berlin

Margot was folding bed-linen with her mother, a white sheet billowing between them like a sail. 'It's good to have you home, Mama.'

'I'm impressed that you kept the place so spick and span – you're going to make someone a nice little homemaker one day.'

Margot's heart-shaped face coloured. She was in her twenties now, yet she did not seem to have changed much from the little girl who curled up on the settee with her books. She appeared to have little interest in men, but at least she had stopped hiding under the table if her exit was trapped when one of her stepfather's guests arrived.

Einstein and Elsa had discussed Margot while in America. Elsa thought her in danger of spinsterhood. 'What a waste of a life,' she had said, 'having no one to care for.'

Ilse, on the other hand, was out with her latest beau, a literary editor and gentleman named Rudolph Kayser. The relationship with Nicolai had revealed itself to be a dalliance, much to Einstein's relief, and he was much more comfortable about Rudolph. He was closer to Ilse's age, for one thing. This time, when Elsa talked about wedding plans, he was prepared to listen.

He was stretched out on the settee, watching Elsa and Margot dealing with the laundry. Elsa had moved on to regaling her daughter with tales of the American standard of living.

'You know I detested the whole thing,' he said. 'I was nothing but a prize exhibit paraded from one arena to the next.'

'Take no notice of him. He loved it.' Elsa looked at her husband. 'You were grinning well enough when they carried you on their shoulders.'

That had happened at the formal welcome in New York, where ten thousand people had gathered outside City Hall to hear the speeches. Einstein had not addressed the crowd and could only just understand the prepared statements, yet the throng had given him an even more rapturous welcome than on the docks. As he left, they surged after him and he was hoisted on to his colleagues' shoulders to be poured into the back of an open automobile. Elsa had to fight her way through to stay with him. Then they set off for a lap of honour around the city.

Unable to contain himself, he stood during the motorcade, as much of a spectator as all of those waving from the sidewalks. He watched them with naked astonishment and they watched him in much the same way. Their attention was unfathomable. They packed the Metropolitan Opera House from floor to rafters.

'Most of them couldn't speak German,' explained Elsa with an incredulous laugh, 'but they came anyway. There was no translator for the evening, and yet they still stood in the aisles to hear him speak. Then in Washington there was that reception at the National Academy of Sciences.'

'What a waste of time that was,' he cut in. 'Taught me a whole new theory of eternity.'

'You loved every minute of it.' Her comment drew a girlish giggle from Margot.

Einstein sank back into the cushions. He had made a tidy sum for himself with his lectures, but the truth was that the Zionists had fallen a long way short of their target. He had learned from Weizmann that they had hoped for four million, but had raised less than one. It was barely enough to start building.

One night in a luxurious hotel suite, Weizmann had asked Einstein if he was really a Zionist.

'I've never made any secret of the fact that I'm an internationalist,' Einstein had replied.

'Would you consider living in Jerusalem?'

'No, I would not.'

'Not even as chancellor of the university?'

A tiny chord sounded inside Einstein, like a church bell in a distant village, but it was too far away to have any real significance. 'Not even as chancellor,' he said.

Weizmann turned to face him. 'Then let us pray that question never comes up.'

'I can only speak as I feel,' said Einstein.

'What was that, Albertle?'

He must have lost himself in the memory and spoken the words aloud. 'Nothing,' he mumbled to his wife.

There was a businesslike rap on the front door. Elsa bustled off, muttering to herself, 'How do they know so quickly when you're back?'

Einstein shrank into the cushions but strained to hear.

'You again,' said Elsa in surprise.

Einstein rolled on to his stomach and peeped round the upholstery. Margot stifled a nervous giggle at his antics.

The stranger raised his hat to Elsa. 'It is, indeed, me again, Frau Einstein. Thank you for remembering me from before your trip.'

'Do you have the money? Remember what I said – no money, no interview.'

Einstein watched dumbfounded as the man reached into his inside pocket, pulled out a sealed envelope and handed it over. Elsa immediately popped the seal and inspected the contents. Her whole demeanour changed. 'Herr Dimitri Marianoff,' she said magnanimously, 'allow me to introduce you to Professor Albert Einstein, father of the relativity theory.'

The man stepped forward, an impish look on his young face. Einstein reluctantly stood up and stuck out his hand. It was shaken vigorously.

Elsa informed them that she would return in an hour, by which time the interview must be concluded.

The young man agreed. He noticed Margot, frozen and staring at him as if he were a burglar. Einstein was about to apologise, but Marianoff spoke first.

'Good afternoon, Fräulein. What a pleasure to be in your company.' He doffed his hat and bowed.

To Einstein's astonishment, Margot curtseyed. 'The pleasure is all mine.'

By the time Elsa came back, Einstein was pacing the hall as quietly as he could.

'What have we here?' she asked.

Einstein held up a finger for her to be quiet. 'They've been discussing Russian theatre. Now he's analysing her handwriting.' He beckoned Elsa to the doorway and they peeped round. Margot was spellbound, staring at their visitor as he talked.

'I didn't know what to do,' said Einstein. 'I felt in the way. Should we stop them?'

Elsa gave him one of her don't-be-ridiculous looks. 'The first male stranger she hasn't run away and hidden from?' She pushed past him into the room.

'Herr Marianoff, would you like to stay for dinner?'

'Yes, please,' said Margot on his behalf.

Inside the lecture room, the Academy looked much the same as before. Nothing had really changed in the audience: there were the familiar grey men with paunches alongside the younger scientists trying to affect confidence. Yet Einstein felt like a child returning to his parents' home and finding that the towering staircase now looked shrunken and all the cupboards were within easy reach. The familiarity of the setting emphasised to him how many months he had been away.

He seemed to be having trouble catching people's attention. Perhaps it was because the presentations were due to start, but no one seemed to be looking in his direction. If someone did make eye contact, it was brief and his smile was ignored. He chose a seat at the back of the hall, suddenly reluctant to force his way to the front.

No one came to sit near him. He doggedly sat through the presentations, but did not really hear a word for the sound of blood pumping in his veins. When the meeting was called to a close, he stood up and left. No one bade him goodnight.

He was still smarting from the slight the next day when he arrived at the university. He grumpily pushed papers round his desk, pausing only when Planck shuffled in, muttering, 'I'm sorry, Albert.'

'So you heard then?'

'Scientifically, of course, it's yours, but politically you've stirred up so much trouble.'

'What's mine?'

'It's just that you've made some bitter enemies.'

'What are you talking about?'

Planck stared at him. 'The Nobel. Oh my goodness, you haven't heard, have you?'

Einstein dropped his pencil to the desk. 'No, I haven't, but let me guess – despite all the nominations, they've awarded it to somebody else.'

Planck grimaced. 'Worse, I'm afraid.'

'Not one of the quantum boys, surely?'

'You must get over your prejudice.'

'Why? The theory's wrong. It needs correcting.'

'Let's put that particular gripe of yours to the side for a moment.'

'All right, but tell me who has won.'

'No one.'

'No one?'

'The committee has declared there will be no 1921 prize for physics.'

It took several moments for the full meaning of Planck's words to sink in. 'So they would rather award nothing than give it to relativity. The idiots.'

Planck tried to strike a conciliatory note. 'The public interest surrounding you is damaging your scientific credentials. Why not live a quieter life? I heard about what happened at the Academy last night.'

'Is it because I'm Jewish or because I visited America?'

'You went to Britain and France as well.'

Einstein lifted his chin. 'Yes, I did. Every German should visit France. They took me out and showed me the battlefields. They're still bare after almost four years, nothing but sludge and dead grey trees. The graves stretch as far as the eye can see. Yes, every German should be made to see that. Do you know, in America they parade their Jewish war heroes? Can you imagine that happening here?'

An urgent knocking on the office door saved Planck from having to reply. Glad of the interruption, he stepped aside as the visitor entered.

'Albert, thank heavens you're back. I came as soon as I heard.' Kurt Blumenfeld was twitching with nervous energy.

Einstein rose from his seat. 'Whatever is the matter?'

'It's that idiot, Rathenau.'

Rathenau waited until they were sitting at the extravagant dining-table before taking in both Einstein and Blumenfeld with a single steely look. 'So, I take it you have come to congratulate me, gentlemen.'

'Is it really wise to accept the post?' asked Einstein, tucking a napkin into his open collar.

'To be appointed Foreign Minister for my country is a great honour. I consider it a reward for the work I have done, overseeing the reconstruction so far. How could I refuse?'

Blumenfeld said, 'Because you will become a target for the far right. How can you not be? A Jew conducting Germany's foreign policy? The idea is laughable.'

'Only if you insist on drawing such an artificial distinction. I am German, descended from Jews. How many times must I repeat this? Working for the good of Germany is the most powerful thing I can do to undermine anti-Semitism.'

'But your policy is not working. Even now the Nazis are recruiting, saying that Jews are parasites on the German economy.'

Rathenau tutted. 'Stop panicking. The National Socialist German Workers' Party is a fringe organisation at best. They're doomed to fade away as economic conditions recover.'

'But more people are becoming sympathisers every day.'

'And who can blame them?' Rathenau said in conciliatory tones. 'They are simple workers who are suffering under ridiculous levels of inflation, struggling to find jobs . . . I plan to negotiate a reduction in the reparations. Everyone – even the French – knows they are excessively punitive. Once the debt eases, the workers will find jobs again and the Nazi party will evaporate. Equilibrium will be restored and I, a German Jew, will have been at the centre of the solution.'

'So, you do admit that your actions reflect on us all.' Blumenfeld leaned back to allow a soup-plate to be placed in front of him.

'In that sense, yes.'

'Yet, in general, you refuse to be identified with the Jewish people.'

'I have not made a secret of my faith. I still worship as a Jew, I still wear a *kippah*, but not in public. I consider one's religion a private matter.'

'You will never be taken as a German, even if your plans succeed. If your plans fail, then you will be vilified as a Jew – and all the rest of us will be made to share the consequences.'

'Gentlemen, gentlemen, let us not argue. The soup will spoil.' He picked up his spoon and began to eat. "Tell us, Albert, what are the American universities like? One hears so much of the place.'

'I was struck mostly by Princeton. The university there had a . . . it had the freshness of an unsmoked pipe.' Einstein lifted his own cutlery. 'Unlike Harvard, where they thought it best to talk to me in French rather than German.'

His host laughed at the absurdity; Blumenfeld sulked.

'Let us hope your message of reconciliation continues to spread,' said Rathenau.

'I managed to thaw out some of the English.' Einstein told how, in preparation for a lecture, he had visited Westminster Abbey with Eddington and laid flowers on Isaac Newton's grave. Despite this gesture, a resounding silence had greeted him as he took the podium at the Royal Society.

He spent the first part of his discourse eulogising the great British scientists whose work he felt humbled to have been given the opportunity to build upon. When the event ended, the floor rose and he received a standing ovation. At that moment, he would have traded those hundred scientists for the tens of thousands of people who had mobbed him in America.

'Mind you, my efforts do not seem to have made me any more popular in Germany.' He recounted the story of his snubbing at the Academy.

'It will take time; we must have patience.' Rathenau dabbed at his lips with a napkin. 'Where next?'

'Switzerland, to see my boys. They're growing so fast – my eldest is at college already. I plan to take him to Florence and show him Galileo's tomb. I seem to be on a tour of my predecessors' graves. Makes me wonder who will visit mine.' Einstein laughed at his own joke. 'Walther, I fear for you. By accepting this post, you really do make yourself a target. There are dangerous, irrational people out there. Perhaps it's better that we Hebrews keep our own counsel, and in that way rise above the pettiness.'

Rathenau spoke with iron calm. 'My decision is final. I will do my best as Foreign Minister. I will live my life as a German and, when the time comes, hopefully a long way off yet, I will die as one. Unlike either of you, I have faith in Germany. Now, gentlemen, let us enjoy the rest of the meal. I pray that the meat is not too tough for you.'

24

Florence, Italy

All was quiet and cool in the basilica of Santa Croce. Footsteps echoing from the tall stones, Einstein and Hans Albert made their way into the interior. Here and there worshippers sat on the narrow pews, mostly women behind widows' veils, as motionless as the statues on the tombs.

Einstein and his son came to a halt. 'This is what I wanted us to see.'

Galileo's sarcophagus was made of ochre marble, with a bust of the great astronomer sitting on top. Bearded, his face lifted to the stars, it seemed as if he were about to raise the telescope in his right hand to his eye. A globe of the Earth and an assortment of books provided a rest for his left arm. Yet with his gaze turned up, he was forever robbed of looking back at those who came to visit.

'How do you pay your respects without praying?' whispered Hans Albert. 'I understand giving thanks for a life and praying for their place in Heaven, but what if you don't believe? What do we wish for then?'

Einstein turned the problem over for a moment. 'We pay our respects by thinking of how much poorer the world would be without their contribution.' He ran his own eyes over the muses flanking the sarcophagus, both portrayed as women in flowing robes: one for astronomy and the other for geometry. 'Galileo was originally buried under the bell tower here in an unmarked grave. He was only given a proper monument and moved inside here during the late 1730s, '37 I seem to remember.'

'Was he the greatest astronomer?'

'His real achievement was his physics, with his studies of motion and his mathematical way of presenting his findings. For me, the greatest astronomer was Johannes Kepler. I simply don't understand why he's not as well remembered as Galileo. Kepler's laws of planetary motion were herculean and proved that Nature could be captured into numbers. He and Galileo set the stage for Newton.'

There was a shushing from the pews.

With a final look at Galileo's upturned stone face, Einstein signalled for his son to follow him and they returned to the daylight.

They wound their way through the shimmering streets, sticking to the shadows wherever possible until they emerged on to the Piazza Uffizi with its overwhelming yellow Palazzo in one corner. With its single thrusting finger of a spire, it was the seat of the Medici rulers in Galileo's time. Drawn by the tempting smell of almond pastries and ground coffee beans, Einstein chose a café nearby, crowded with mid-morning customers.

His son looked relaxed in a lightweight grey suit, with a face that was full and round. Spectacles sat proudly on his nose, adding to his air of confidence. College was clearly suiting him.

'You want ice cream?' asked Einstein.

The young man gave him a look of amusement rather than disapproval. 'Tete still likes ice cream, Papa. I prefer coffee.'

Suitably admonished, Einstein ordered the drinks and raised his face to the warmth. 'I want to say that I think that perhaps you made the right choice of career.'

'You do?'

'Yes. I didn't think so at first but engineering is a solid choice, a good path. My father and uncle were both engineers.

You won't have to go searching for some four-leafed clover to prove your name. Just build us some solid bridges. I think you're wise.'

The drinks arrived and his son swigged the coffee with a broad smile. 'I read about physics, but I don't understand a lot of the modern ideas.'

'Neither do I, but no one believes me when I tell them that the quantum theory is flawed. The laws of physics should determine precisely what happens, not leave things to chance. Light spitting itself out of atoms at will – rubbish. Utter rubbish. Galileo left nothing to chance in his science. An effect, must always have a cause. Quantum theory cannot be correct, but no one listens. They just bury their heads in the sand.' Einstein's voice had risen and was drawing looks. 'Sorry, son, I didn't mean to embarrass you.'

Hans Albert looked surprised. 'You don't, not any more.'

'Any more?'

'Sorry, Papa. Mama talks a lot about you, tells me about you.'

'I dare say your mother knows me better than anyone else.' The thought brought both warmth and insecurity. 'She knew me before you were born, watched me change into what I am today.' He paused. 'We can't help the way we change, even if it's hard on those around us.'

'I know you don't necessarily mean the unpleasant things you say. You're distracted, thinking about your work. Mama says that you can be hurt by things we say, well, things I say. I'm sorry for that.'

The square, for all its activity, drained away from Einstein's consciousness.

'Have I upset you, Papa?'

Something was tugging at his heart. 'Not at all. I should be the one apologising to you.' He recalled Mileva in her wedding frock, high-necked with a tiny pinched waist, and the delicious look in her eyes. 'I think I should visit your mama when I take you home. Do you?'

'Please, Papa. It would make her so happy. Tete, too.'

Eduard was now seventeen, stocky and awkward. He hung on to his father in the hallway the way he used to as a boy, seemingly unaware of his burgeoning strength.

'Tete, be gentle,' said Mileva, straightening her blouse at the sight of her former husband.

'No, it's all right,' said Einstein, finally released.

Mileva's cheeks were fuller again, her lips pleasantly coloured. 'Did you have a nice time?' she asked her elder son.

'Oh, yes. Splendid. We saw Galileo's tomb.'

'Don't worry,' said Einstein. 'He's still going to be an engineer.'

They laughed more loudly than the feeble joke deserved. They all shuffled a little. He thought about hanging his hat on the coat-stand, then decided it was too forward. Besides, Hans Albert was willing her with his eyes.

Finally her eyes sparked, and with a slight tremor in her voice she invited Einstein to stay for tea.

'I would like that very much.'

'Tete,' said Mileva, 'hang up your father's coat.'

He all but tore the garment from Einstein's shoulders.

After the tea and some idle chatter – although there was a point when Einstein became fixated on revising the Fibonacci sequence with Tete – the family relaxed.

'That was lovely, thank you, Mama.' Hans Albert stood. 'Come along, Tete, I want to tell you all about Florence.'

His brother looked delighted and slipped from the table. Einstein saw the triumph on his elder son's face as he led the boy away, leaving his parents alone.

They sat for a moment. Mileva ran her finger along the edge of her collar. Einstein eyed the clutter in the place.

'I know,' said Mileva, 'I should tidy up.'

'I like it, feels homely.'

The comment embarrassed them both, neither sure what lay behind it. Was it just politeness, or something deeper?

'How bad is it in Germany?' she asked. 'We read things in the papers and we hear stories.'

He was grateful for the change of subject. 'The currency has collapsed. They've printed so much money to try to meet the war reparations it's now virtually worthless.'

'Yet you always meet your payments to us.'

'My foreign lectures are paid in dollars; I keep it in bank accounts outside Germany, where it's safe. Otherwise I'd be on the streets.'

'Just you? You never mention . . .' Her voice trailed off.

'Elsa,' he said gently, 'I don't want to upset you.'

'That's very kind of you.'

'It's not a marriage like we had together.'

Mileva made to get up from the table. He continued, anxious to keep her attention. 'She's more of a governess than a wife,' he blurted.

'Don't, Albert.' Mileva's hand trembled as she collected the tea-things. He cursed his clumsiness. After a few minutes he followed her to the kitchen, pausing at the door as she began filling the sink.

'I'm worried about Tete,' he said tentatively.

'He's not like Albert, that's for sure.' Mileva shrugged.

'For seventeen, he seems . . .' How could he say 'retarded' to the boy's mother? 'Has he been sick lately?'

'Small things, nothing major. It's more his emotional state; the least little thing can trigger one of his moods. He's all ups and downs.' There was concern in her voice.

'I've always worried about him, and soon he'll be at university. Can he look after himself?'

'He'll stay here and go to university. He's thinking of studying psychoanalysis.'

'What a waste of time.' The words were out of his mouth before he knew it.

Mileva eyed him over her shoulder. 'Same old Albert, eh?'

'Sorry.' Flummoxed, he reached for a tea-towel and set to on the pile of crockery Mileva was building next to the sink.

'Does he still play?'

She continued to flourish the dishcloth. 'Piano? Oh yes. It's one of the few things that he will always settle to. We have a musical evening planned with the Hurwitz family tonight.'

'You still do that?'

'Of course. Come along if you like. You could borrow a violin. We could drop you off at your hotel on the way home.'

'I haven't booked one.' His heart accelerated. 'I was wondering about the little room upstairs.'

She stopped washing up and turned to face him, hands and dishcloth dripping. They were standing closer than they had for years. There was still a wall between them, but it was a thinner one.

'I'll not inconvenience you, Mileva. We're a family. I may have sometimes forgotten that.'

'The boys would like it if you stayed.'

With a nod, he picked up another plate and slid the towel across it.

'And so would I,' he heard her say in a voice that was barely a whisper.

25

Berlin

There had been plenty of laughter at the musical evening. In the relaxed atmosphere Tete had stolen the show, reducing his father to tears with the sensitivity of his playing. The next day, Mileva and the boys had accompanied Einstein to the station, and he had tried to hide the leaden movement of his feet. Amid the whistles and the steam, he had stood close to her. 'You live in a beautiful country, you have two beautiful sons, and, best of all, you don't have me getting in the way.'

She tried to smile at his bravado. 'I miss you, Albert, still.'

'I know,' he said tenderly, 'I don't understand why, but I know.' His eyes had brimmed with tears as he hurried to the carriage.

Now back in his attic study, Einstein was still musing over the visit and his tangled emotions. Mileva was the mother of his children; Elsa was his wife. He was bound to both of them in ways that he had failed to grasp before.

A sharp knock on his office door brought him back to the present. He shot a poisonous look at Elsa, who appeared in the crack of the door. An admonishment for disturbing him was on his lips until he saw how pale she was.

'Albertle, it's the police. They've refused tea.'

He found the two officers in the sitting room. They stood up stiffly when he entered.

'What can I do for you, gentlemen?'

'It's Walther Rathenau.'

Einstein sat down heavily, guessing what was to come.

The officers explained quietly and methodically how the Foreign Minister had been in his car that morning as usual, being driven to work. According to witnesses, his car had slowed at a crossroads. As it did so, a vehicle carrying four young men, each identically dressed in leather coats and driving caps, pulled up alongside. One of the occupants drew a gun and took aim. The Foreign Minister had stared at his killer as five shots rang out.

'He had been advised to vary his daily routine,' said the officer. 'We think they must have been watching his house for a while.'

Einstein berated himself for not doing more. He had known that this was the only way it could end, yet he had allowed himself to be fobbed off by Rathenau at their supper. Damn the man and his obstinacy! And damn him for his craving for power, which had ultimately been his undoing.

Einstein thought of Rathenau's bullet-ridden body, probably not yet fully cold, lying on some slab somewhere, awaiting preparation for burial.

'Have you caught those responsible?'

'Yes, sir, within the last hour. They were not very sophisticated. Seems as if they didn't realise we'd come after them so quickly, if at all. We caught them relaxing back in their apartment.'

The idea of their failing to understand the deed was punishable shocked him more than the crime itself.

'There's something else.' The chief officer's face was grave. 'A list of names found at the address. Your name was on it.'

Elsa gasped from the doorway.

'A list of targets?' Einstein asked.

'We believe so. Herr Rathenau was top of the list.'

Einstein thought better than to ask where own name had been on the list. The fact that the officers were here so quickly told him all he needed to know.

'Is there anywhere you can go? Away from Berlin, I mean,' said the officer.

'The Japanese lecture tour,' declared Elsa, stepping closer to them.

'We're in the preliminary stages only, Elsa. It will take months. Let's not be too hasty.'

'Then we must sell. Sell up and leave Berlin at once,' she said, urgency in her voice. 'There's that company in Kiel, they'll have you in an instant.'

Einstein drew a deep breath, thinking about the compass he had designed for them during the war. 'Elsa, let us stay calm.'

'You must be cautious, Herr Einstein,' said the policeman, eliciting a weak smile from Elsa. 'We would urge you not to make any public appearances. When you do go out, vary your routine. You cannot afford to take any risks.'

Einstein glanced at his wife, intending to give her a reassuring wink, but her look of naked fear froze him to the core.

'Elsa, cross over.' Einstein did not wait for her to react. With a quick check for traffic, he took her elbow and stepped out into the glistening road.

Just ahead of them a group of jostling women were crowded around a factory gate. Wages were now distributed twice a day, and this was the crowd for the lunchtime handout. Workers' names were called out from behind iron railings and the bulging wage packets were handed through to their respective owners. The women snatched the money and fled, desperate to buy anything that was edible before the inevitable afternoon price hike.

Food was scarce and prices so astronomical that it was wise to avoid crowds. There were gaunt faces and hungry people everywhere; the papers were full of recriminations. Public beatings and alleyway murders were increasing, the victims mostly Jewish.

Einstein tried to hide the two small bags of groceries he was carrying and quickened his step. 'I told you not to wear a fur today. It attracts too much attention.' He had suggested a walk in the park, yet somehow it had turned into a wholly unpleasant shopping trip. 'We should have had the groceries delivered.'

'We can't afford to any more, and there's no guarantee they would arrive anyway. What is happening to this city? Have you sent the letter?'

Einstein marched on. 'I've torn it up.'

She looked at him aghast. 'How could you?'

'I don't want to be an engineer.'

'But we discussed it after the policemen left.'

'You were scared. I wanted to calm you.'

He had to admit that the thought was not entirely without merit. Up on the Baltic coast it would be far enough away to keep a low profile. There would be peace and quiet, not to mention the chance of learning to sail. Yet in his dreams he kept finding himself in America, walking the leafy paths across the Princeton campus.

Elsa was tugging on an ostentatious earring. 'What if they target the girls?'

'They won't.' The words sounded lame, even to him. 'And you know how much you like it in Berlin. You like the life here, the girls like it here. We can't take the girls away now they've found gentlemen.'

'Don't blame me for this, Albertle, or them.'

'I'm not. It's just that Berlin is the world capital of physics. I can't leave. Anywhere else is a backwater, and will be used by my enemies to diminish my theory. I have to be here to remain credible.'

'At the expense of remaining alive?' There was a stark challenge in her eyes.

Einstein's mind flicked back to Rathenau and his stubborn quest for power. A wave of uncertainty washed over him.

'It strikes me you spend more time on politics than physics these days,' snapped Elsa.

He tutted loudly. 'I'm working on new ideas to fix the quantum theory. The Good Lord does not deal in chance.'

'The Good Lord now? Oh, that's a change! What's come over you?'

'I don't deny a higher power, Elsa,' he said tartly, 'just the folly of human worship. Why glorify these stories of the past? Why yearn for them and worship them? It shows the people have completely forgotten how to live for the future.'

'Don't sidetrack me, Albertle. You know how I feel about your safety.'

'And don't try to make out this is just about me.'

They arrived at the apartment block. Elsa unlocked the door and they went into the hall. He rested the bags on a ledge to flex his arms before carrying them up the stairs. He took the opportunity to face Elsa. 'Let us live quietly until Japan. That trip will take months, and by the time we return all of this will have blown over.'

'Do you really think so?' Her face wanted to believe him.

'I do.' He drew her into him, enclosing his arms around her soft torso. This way, she could not see the doubt in his eyes.

The packing trunks were open and looking more unpacked than packed. Dresses, hats and suits were strewn everywhere. Half a dozen stiff collars poked upwards like a bunch of bananas.

'We'll never get it all in,' said Elsa. 'We need another case.'

'Just unpack some things. I don't need all those suits. One will do,' said Einstein.

She gave him a withering look. 'I let you keep your hair like that. It's the only compromise I will make.'

Only because you can't manhandle me to a barber, thought Einstein. He ran a hand through the wiry mass on his head.

A telegram arrived.

Elsa threw her fox fur into a pile of dresses. 'Is there no peace!'

Einstein noted the return address. 'It's from Stockholm,' he announced flatly.

Elsa's mood transformed at once. 'Wait, wait,' she said, 'don't open it yet.' She raced to the bedroom and returned moments later, having applied lipstick. 'All right, open it.'

'It's probably nothing,' he said.

'They're not going to write to you if you haven't won.'

He broke the seal and read slowly: 'It will probably be very desirable for you to come to Stockholm in December, and if you are then in Japan that will be impossible.'

'You've done it!' She all but jumped for joy. 'They're finally going to give it to you. You're going to be a Nobel laureate.'

'It doesn't say that.'

'Yes, it does. For a genius, you can be so obtuse.'

Einstein had dreamed of this moment every year since 1905 and the first paper on relativity, even when he had feigned indifference to those around him. He had imagined opening the telegram and getting the news – the recognition of his peers – but now it had really happened, he felt none of the elation he had expected.

'We can reschedule the lecture tour,' said Elsa. 'Once you're a Nobel laureate, we'll be able to put your fees up again.'

'Elsa, wait.'

She studied him. 'What's wrong?'

'I'm not going to Stockholm. They've prevaricated for too many years for me to take them seriously any more'

'You're going to refuse?' Her hands started trembling.

He took hold of them and held them tightly. 'No, no, I'm not going to do anything as melodramatic as that. I've promised the prize money to Mileva – she deserves it – and we're making enough from the lectures now. But I find myself unmoved. They can give me the prize or not, but I'm going to Japan. Coming?'

26

Mount Wilson, California

From the base of the Californian mountain, where the road forked and the winding upward trail began, all Lemaître could see was pine trees and sheer climbs. The vertical spires of vegetation pointed to the sky, hinting at the mountain's true purpose.

The road was dusty with compacted earth baked into powder by the sun, and Lemaître fanned himself in the back of the cart as the mule and its driver waited with the same dreary expression on their faces.

They were parked beside a wooden guardhouse, within which a uniformed man drummed his fingers, staring at a telephone.

'Gotta wait for them to lower the gate at the top,' he explained without looking up. 'Road's single-file and there's no place to pass.' The phone rang and the guard answered, then trudged out to lift the barrier.

The mule strained against its harness and the cart rocked into action. Lemaître stared out over the valley. He had still not grown accustomed to there being so much natural land in America. He had travelled through vast tracts of forest and open plains, deserts and mountain ranges. The world was so staggeringly large compared to the individual. The less significant he felt, the more at peace he became.

By the time an hour had passed, it seemed as though the mule had wound the cart to the top of the world. The blue sky pressed down from above, rich with the promise of a clear night, and the observatory finally came into view.

Lemaître gaped at the size of the domes, rising in tall columns above the trees. Catching the sunlight, the white structures were almost too bright to look at; powerful emotions gathered in his chest, took him by surprise. He was here. The largest telescopes in the world: the means to reveal God's realm as never before was at his fingertips.

Edwin Hubble was silhouetted against the window, his shoulders so broad they impeded the flow of light into his spartan office. His back was turned and he was holding up a photographic plate for inspection.

Lemaître placed his valise on the floorboards and waited. After a time he cleared his throat.

Hubble turned. His eyes, set in a square block of a head, channelled power more like a politician's than a scientist's. Lemaître's thoughts flitted back to Harvard and the scurrying mouse of Shapley.

So, these were the two great rivals.

Hubble took a step forward and Lemaître had to force himself to stand his ground.

'By jove!' said the giant.

Lemaître laughed at the mock-English accent and the bearish astronomer looked horrified at his reaction.

'I'm Georges Lemaître,' he gabbled. 'I wrote to you.'

'By jove!' said Hubble again. He produced a pipe and clamped it between his teeth with an audible bite. He cast the photographic plate aside on the desk.

Lemaître winced at the careless way he discarded it. At Harvard, photographs were treated as fragile treasures, and most people had taken to wearing cotton gloves when handling them.

'Don't worry about that. I can always take another,' said Hubble casually. 'What did you want to see me about?'

Lemaître tried to order his thoughts. 'The spiral nebulae – I'm interested in learning more.'

'Aren't we all, old chap.' Hubble looked completely bored.

It dawned on Lemaître that the accent was not being put on for comic effect. Desperate for something to say, he said, 'I came here from Harvard.'

'Harvard.' Hubble's eyes narrowed. 'How's Shapley?'

'He's well, thank you. He's as intrigued as the rest of us are about your work.'

'So he should be, the Cepheid work is his technique. If he'd stayed here, I think he would have worked on the spirals rather than me.' He pulled the waggling pipe from his mouth. 'Does he accept that Andromeda is an island universe?'

'Another galaxy? Yes, I think he's coming to terms with it.'

Hubble's thin lips cracked into a smile. 'I remember your letter now. You're a theorist, are you not?'

Lemaître nodded, somewhat apologetically.

'Well, well. Who'd have thought it?' Hubble cocked his head to complete his guest's confusion. He thrust the pipe into his tweed jacket's breast pocket and grabbed the glass plate back from his desk. 'Can't chat now, old chum. Come back this evening. I think we'll be able to show you something. I'm on the 100-inch tonight. Show you what real astronomy's all about.' Hubble turned his back and lifted the plate to the window again. Lemaître could see greasy finger-prints all over it. He picked up his small case and departed. He paused, after closing the door as quietly as possible, to shake his head.

'I've seen that look before.'

The voice's owner was a trim woman who walked in neat steps, as if choreographed. Her outline had something of the starlet about it. Despite squinting, Lemaître could still not discern her face, but he got the distinct impression that she was judging him.

'I'm not convinced I've made a very good first impression,' he said ruefully.

'Are you French?'

'Belgian.'

She stepped out of the light. She was young and lean, not quite beautiful, but confident. 'Take no notice of him, Abbé. It's his way.' Her words were consolatory yet there was pride in her voice, as if she enjoyed the fact that Hubble put people off balance.

'I'm impressed you know my title in French.'

She looked perplexed. 'Why? You use it on your letters, Abbé Georges Lemaître. And my husband and I were married as Catholics. I'm Grace, by the way.'

She thrust out her hand and they shook. Her grip was weaker than he had expected from her behaviour.

'I'm intrigued to know how you managed to slip past me earlier,' she said.

'By accident, I assure you.' He turned sideways and laid a hand on his paunch. 'I don't think I'm the slipping-past type.'

Grace fixed him with a stern look. 'Why, Abbé, if I didn't know better, I'd say you were flirting with me.'

'No!' He crumpled under the accusation.

Grace laughed, delighting in her triumph over him. 'Have you been shown to the monastery?'

Lemaître frowned.

'No, silly,' she said, flapping the air. 'It's what the astronomers call the night quarters. No women allowed. Except me, of course.'

He followed her meekly, carrying his luggage. The 'monastery' was a wooden, single-storey curved wing of a building. She walked into the reception area and scanned a ledger, then plucked a key from a hook on the board. His name was written on a fresh cardboard tab fixed to the finger-long metal key.

'This way,' she said, heading down a dim corridor.

He followed. The floorboards creaked underneath the worn rugs.

She unlocked a room and swung the door open for him. 'Dinner is at five. Gives it time to settle. Don't eat too much,' she glanced at his stomach, 'Or you'll never stay awake for the observing. Night lunch is at midnight.'

'Thank you, I did get used to the observing routine at Harvard.'

'Oh, did you? Well, you'll find things different here.'

'I didn't mean it like that,' he began.

'Get some rest, Abbé.' She began to walk back towards the reception area.

'Miss Grace,' he called, 'forgive me, I don't know what your position is here? Are you the secretary?'

Her acid laugh rang out. 'Oh, you are funny, Abbé. Secretary, indeed.' She fixed him with a stare. 'I'm Hubble's wife.'

He closed the door and all but threw himself on the bed.

The dining-room was in the other wing of the monastery. Lemaître had only to follow the smell of onions to find it. He disliked the way the floorboards announced his arrival; he might as well have been wearing a cowbell. He exchanged pleasantries with the others in the room and found his place by the name tag on the table. Considering it was little more than a hut, the place was pleasantly furnished, and the view over the surrounding valleys and mountains was superb.

Hubble swaggered in, wearing dark brown jodhpurs and thick woollen socks pulled up to his knees. Everyone else ignored the bizarre attire so Lemaître did too, discovering later that it was Hubble's usual observing gear.

The giant man filled the seat at the head of the table and cast a monarchical eye over his guests. 'By jove, Dr Lemaître, I need a telescope just to see you. What on Earth are you doing away down there?'

'I sat at my name tag.'

'No, no, no. Someone's been larking around. This won't do at all. Spencer, swap positions.'

An elderly man looked up from Hubble's immediate right. 'But I'm principal observer on the 60-inch tonight.'

'Be polite,' commanded Hubble.

The displaced astronomer flapped his napkin petulantly on to the table and got up. Acutely embarrassed, Lemaître relinquished his seat and made for Hubble's side.

'Strict seating order, you see, depending on who's using what telescope,' Hubble said, sounding more clipped than before. 'Doesn't mean we can't change it though, does it, Georges?'

Lemaître would have preferred to stay where he was. He could feel the scrutiny of the others.

'There are enough games being played with the seating already,' grumbled Spencer with a sidelong stare at Grace.

She stared back defiantly.

When the meal arrived Lemaître served himself sparingly, noting that the others did not seem so restrained. Their talk centred on the observation of the universe, nothing else mattered. The astronomers gossiped about spectroscopes misbehaving, or the telescopes wandering from their targets, as if the pieces of equipment had minds of their own. If it had not been for his time at Harvard, he would have been completely at sea.

'Tell me, how is Europe recovering from the war?' Grace cut across the chatter.

'Slowly, I'm afraid. The destruction of the land and the buildings . . .' He shook his head at the memories of the rubble. 'The Germans indiscriminately destroyed buildings and people as a means of intimidation. There is nowhere one can really go in Belgium without finding a reminder of the occupation.'

'Yes, we felt so sorry for you all over there. My husband fought, you know. Volunteered as soon as America entered the war.'

'In 1917?' Lemaître asked.

'That's right, cut short my thesis to go and fight,' said Hubble. 'Couldn't leave you boys to the mercies of the Kaiser, could I?'

'What about you, Abbé?' asked Grace. 'You were too young for the fighting, I guess. Must have been grateful when the liberation came?'

Lemaître cocked his head. 'I volunteered with my brother at the outbreak and served for four years.'

'Four years?' Hubble sounded incredulous.

'Yes. I started in the infantry and then I was transferred into the artillery. Fought at Diksmuide, Ypres, among others.'

'My husband was quite the hero, you know. Knocked unconscious by a bomb blast, he woke up in hospital, got dressed straight away and rejoined his unit.'

'Thank you, Grace,' said Hubble curtly. 'My war record is of no interest tonight.'

Grace looked perplexed, her eyes darting between her husband and Lemaître.

'Now then, gentlemen, back to astronomy. What's the seeing like tonight? Didn't look too soupy on the way over; I'd say we could be in for a good one.'

Dusk was well advanced when Hubble took Lemaître through the shadowy paths to the domes. The blinds on all the buildings were tight against their windowframes to prevent stray light fouling the observations. With no illumination, the buildings could have been mistaken for a ghost town. Now and again the silhouette of another astronomer would cross their path, and a muttered greeting would be all that proved it was a person, not a wraith.

As they walked, Hubble seemed to relax; he lowered his voice and most of the feigned English slipped away. 'You must excuse my wife if she makes the occasional lapse of judgement. It's all about appearances here, you know. I blame Hollywood myself, but that's the society we live in: one of self-promotion as a means of advancement. In her efforts, God bless her, Grace tends to put me on something of a pedestal. I'm sure you understand.'

Lemaître did not, but he was willing to accept Hubble's word for it. 'The fairer sex remains a mystery to me.'

'You must have seen a lot of action.' Hubble spoke in the same matter-of-fact way that Belgian fathers who had been at home addressed their sons.

'More than enough.'

'The truth is, I hardly came under fire . . . but at least I went. *Others* simply stayed at home.'

Lemaître looked round, struck by the emphasis. 'You mean Professor Shapley?

Hubble nodded gravely.

'He was a conscientious objector?'

'Conscientious slacker, if you ask me. But you, you fought for four years.' He whistled aloud. They were at the towering dome now and Hubble paused. 'I salute you, Dr Lemaître. You're a brave man. Now, let me show you something truly astounding.'

They stepped into the building and on to a dim staircase. The lights were red to preserve their eyes' adaptation to the dark. It felt as though they were entering another world. Hubble's face filled with pride as he swung open the top door and Lemaître caught his first sight of the 100-inch telescope.

'By jove!' the priest breathed.

'Quite so,' chuckled Hubble.

Everything was in shadow, the lights already out for the night. The only illumination came from the open shutters in the dome's roof, where the Milky Way was splashed across the darkness. In the starlight Lemaître could make out enough to know it was like being in a cathedral transept. Instead of stone, the curving arch of the roof was a skeleton of metal bones, skinned with sheets of aluminium. When they spoke, their voices echoed from odd angles.

Night assistants were readying the telescope, all in matching suits with heavy scarves and flat caps. Hubble ordered them around in tones that Lemaître had not heard since the battlefield, then led him to the base of the telescope.

The 100 inches of the telescope's name referred to the diameter of the mirror, and only when Lemaître was standing under it did he realise how big 100 inches really was. The mirror was two and a half metres in width, held at the base of the telescope's enormous tube by a great metal cradle.

'Two tons of fused glass,' said Hubble, patting the metal casing. 'Took a year to cool it down so it wouldn't crack.'

Lemaître stared dumbfounded. His eyes traced the telescope's tube, which stretched upwards as tall as a house.

'First target, sir?' called one of the assistants.

'Not just yet, I'd like to show our guest M31 first.'

'It's a bit low.'

'Only a visual inspection, we don't need photographic conditions.'

Lemaître followed as Hubble stepped assuredly in the darkness and they ended up on a gantry. From there they watched as the assistants worked a noisy control panel that fired motors to guide the telescope into place.

The great metal beast moved from bolt upright to a recumbent position and the workers manually swung a platform close to its side, where an eyepiece was fixed into a metal plate. They continued to nudge the telescope for a few moments until they called out 'Ready'.

Hubble led the way and checked the view. 'There you go, old boy.' He pointed to a seat that had been cemented to the edge of the platform. Lemaître sat down gingerly, legs dangling over the edge.

He leaned into the eyepiece. The view was dark but textured and swam around as he adjusted his position. He relaxed his shoulders and, as he had learned, waited for his brain to catch up with his vision. When it did, the Andromeda galaxy resolved itself into magnificent life. It was a smudge, an indistinct pale oyster that looked as if it might disintegrate if he breathed out too quickly.

'The light that's striking your eyes started its journey almost a million years ago,' said Hubble, 'before any human being was alive. The galaxy you're looking at tonight is how it appeared a million years ago, not as it exists today.'

Lemaître shivered with excitement. 'Amazing.'

'Quite a thought, isn't it?' agreed Hubble.

'Have you read the H.G. Wells story about the time machine?'

'I don't really go for such works.'

'It concerns a man who builds a machine to take him into the future. It strikes me, Dr Hubble, that your telescope allows you to see into the past.'

Hubble made a contented sound. 'Never really thought about it like that, but I suppose it does.'

'What, I wonder, does this galaxy look like today?'

'I can tell you how to find out.'

Lemaître looked round in surprise.

'Just sit there patiently for a million years, old chap. The light will have arrived by then.'

Hubble began his scheduled observing shortly after. An assistant manoeuvred the telescope into a different direction, this time setting the base close to the floor so that the great cylinder pointed closer to the zenith. Then the dome filled with an almighty rumbling as its motors moved it round to put the opening in front of the telescope.

Hubble grabbed a battered bentwood chair from near the wall and positioned it next to the eyepiece. He loaded a photographic plate into the camera slot and changed the eyepiece for one with a crosshair. He lit his pipe and settled in. For an hour he watched as the photograph collected light and slowly built up the picture, constantly checking the way the telescope was tracking, correcting for any drift in alignment.

One assistant showed Lemaître how to control the dome, edging it round as the telescope tracked to make sure it provided an unfettered view of the heavens. It gave him something to take his mind off the cold.

During a break Hubble came sauntering over, rolling his shoulders and drawing on his pipe, lighting his face with an orange glow every time he took a lungful of smoke.

'Now you know how we track down the Cepheids in these distant galaxies. Patience and careful guiding. The pictures of the stars they take at Harvard are child's play compared to the galaxies.'

'And what, if I may ask, do you make of the spirals and their astounding velocities?' Lemaître added one of Eddington's crowd-pleasing witticisms: 'They shun us like the plague.'

Hubble rubbed his chin. 'I must admit to being sceptical. It still seems extraordinary to me that such structures could be accelerated to such vast speeds. How could that possibly be true?'

'How indeed?' mused Lemaître, his mind brimming with possibilities.

'Do you know something you're not telling me?'

'I thought you hated theory.'

Hubble's face coloured briefly with the demonic glow of the pipe. 'I do, but it doesn't mean I'm not curious. Observations have a purity about them, whereas theories are always a matter of interpretation.'

Lemaître filled the dome with a belly-laugh. 'I think of it the opposite way round.'

'Come along, tell me what you know.'

Lemaître grinned. 'Have you heard of the de Sitter effect?'

27

Brussels, Belgium

1927

Max Planck's baggy eyes were duller than Einstein could ever remember them. Although the old master was nearly seventy now, it was not that long ago that Einstein had seen them blazing with a steel-blue challenge. The voice was diminished, too. 'Why must you insist on derailing this conference?'

'Because it threatens to derail physics,' said Einstein unrepentantly.

'You're behaving like a curmudgeon. You contest every speaker even when they are delivering peer-reviewed and published work.'

'The quantum theory is weak, yet you all discuss it as if it were proven.'

'It is proven. It explains the structure of the hydrogen atom and why it absorbs only certain wavelengths of light perfectly.' Planck sounded weary.

Einstein formed his words carefully: 'It cannot be perfect if it relies on probability. How many times must I say this?'

They were standing to one side of the tearoom. Around them the invited physicists bubbled with conversation but did not approach. Outside in the autumn mists of Leopold Park lesser scientists waited, having made the trip to Brussels just to buttonhole the invitees as they were leaving.

'Your objections are not scientific,' said Planck. 'This is about your hatred for German achievement, but remember this. The world has excluded us from experimental projects, theory is all we have left, and we have built a way of understanding the sub-microscopic world that is the envy of other countries. Now the Danes have taken up our ideas, and you detest that success. Honestly, you can be as poisonous as the rest of the world.'

'If you believe that, we can no longer be friends.' Einstein crossed his arms, hoping to disguise his embarrassment at the *schadenfreude* he relished whenever German nationalism was damaged.

'These men are brilliant,' Planck continued, indicating the gathered scientists. 'Bohr, in particular.'

The man in question was standing across the room, animatedly talking over the top of his teacup. He was just six years younger than Einstein, yet the gangling Dane still had the aura of youth about him. His hair was slicked back, as was becoming the fashion. Einstein had tried wearing pomade once at Elsa's request and hated it so much he had refused to leave the apartment until it had been washed off.

'They're young and impulsive. They jump to conclusions without thinking about the consequences,' said Einstein.

Planck pushed his round glasses up his nose. 'They can't understand why you're being so obstructive, especially because you're one of the fathers of quantum theory. This whole conference is about the consequences of these new rules. Think back to when I first talked about light as packets of energy. I intended the notion only as a mathematical trick to get the right answer, but you convinced us all that light really could be photons, little particles of light flying around instead of rays of light. It's what got you your Nobel Prize.'

'Don't remind me of that.' The mere mention of it caused him to shudder. When he had found the official letter waiting on his return from Japan, he had read with dismay that he was to be given the deferred 1921 award after all, but not for general relativity. The letter informed him it was 'for his services to theoretical physics, and especially for his discovery of the law of the photoelectric effect'. If that had not been bad enough, the letter went on to make it clear that the award was being given 'without taking into account the value that will be accorded your relativity and gravitation theories after these are confirmed in the future'.

Einstein had dropped the letter. 'Confirmed in the future! Does Eddington count for nothing with these idiots?'

His annoyance had risen still more when he discovered that at the awards ceremony, which had taken place while he was in the Far East, the chairman of the awards committee had dismissed relativity as a branch of epistemology and therefore of interest only to philosophers. Incensed by the insult, when Einstein had turned up to give his belated laureate lecture the following summer, he had delivered it exclusively about general relativity.

He clattered his cup and saucer on to a nearby table. 'Yes, Max, I did propose the photon of light, but I also pointed out the weakness of the approach – the way it leaves the time and direction of the emission of the light to chance – but no one has heeded the warning.'

Planck looked suddenly exhausted. 'Oh, Albert, you talk in such dramatic terms.'

'I will not have absurdities destroy physics.'

'Even if the universe itself is absurd?'

Einstein scowled. 'Particularly then.'

The delegates returned to the main room and sat around a horseshoe-shaped table arrangement, with the open end facing a blackboard. Einstein took his seat as Bohr called the meeting to order with the informal air of a bumbling schoolmaster.

'This year,' said the Dane, 'a quite remarkable piece of work was published by our next speaker, Werner Heisenberg. I say, without doubt, that this is one of the presentations we have all been waiting to hear: Uncertainty as a Fundamental Constituent of Nature.'

Heisenberg marched to the front amid scattered applause. Unlike Bohr, who feigned youthfulness, Heisenberg really was in his mid-twenties and possessed the overt self-confidence that Einstein realised he himself had once displayed.

The young German placed his notes on the lectern, checked that there was chalk at the board and straightened his jacket. He was soon into the mathematics, filling the board with the runic symbols of science.

Einstein knew he should be impressed, both with the assurance of Heisenberg's delivery and the robustness of the calculations, yet he experienced a visceral revulsion. He could see what none of the others could, with their wide eyes and enthusiastic nods of approval: the elegance of the mathematics was a Lorelei, and it was luring them all to destruction.

Heisenberg's delivery became even more emphatic as he reached his conclusions. Writing his final equation on the board, he underlined it and punched such an emphatic full stop that the chalk split in two.

'The mathematics can mean only one thing, that one can never know precisely both the position and momentum of a particle. The more precisely we measure one quantity, the less precisely we can determine the other. This is not a question of developing more precise instruments or measurement techniques; this is inherent. Uncertainty is a fundamental constituent of Nature.'

The physicists went into rapture, clapping their hands together and talking excitedly to each other. Heisenberg soaked it up, virtually standing to attention at the front.

Bohr rose from his seat, grinning from ear to ear. 'Any questions?'

Einstein let the sycophancy die down before raising his own hand. He did not wait for Bohr to acknowledge his request. 'Physics is about determining cause and effect. By weaving randomness into the fabric of reality, you lose that battle. Physics will no longer be about determining why things happen. You will destroy our subject if you cling to this line of reasoning.' There was plenty of muttering at that. 'There must be an objective reality from which all of the observable phenomena spring . . .'

'Like Plato's shadows on a cave wall?' challenged Heisenberg. 'What good does such thinking do us? We are physicists, not philosophers. We work with what we can see and what we can measure, and we test our theories by predicting what more we can see and measure. We must allow nothing into theory that cannot be tested. The uncertainty principle can be tested because no experiment will ever give both quantities precisely.'

Einstein made a dismissive sound. 'I cannot hold with this new fashion for believing only in the things that can be measured.'

There was a change in the room, like water turning to ice.

'But you're the architect of that approach,' stuttered Heisenberg, his confidence crumbling into incredulity.

Einstein moved his shoulders in a parody of a French shrug. 'A good joke shouldn't be repeated too often.'

Bohr spoke immediately, his tone firm. 'No, you can't fob us off like the reporters. This needs an answer. In 1905, while all other physicists were labouring under a belief in the ether, you said that if the ether could not be observed, it could not form part of a theory. Without it, you went on to derive special relativity, and now you tell us that such an approach is worthless?'

Einstein leaned back to try to hide a twinge of embarrassment. 'It's possible that I did use that kind of reasoning.' The room stirred at the admission. 'But it's nonsense all the same.'

Bohr wagged a finger. 'The mistake you make is to think that physics is about finding out what Nature *is*. But physics must concern itself only with what it can *say* about Nature. Scientists are not about the discovery of truth, we leave that to the Church.'

There was a ripple of laughter. Einstein waited for a beat. 'Have your quantum theory if you want, but don't believe it's the final word. It's a stepping-stone to a unified theory where certainty will exist whether we can observe it or not.'

'And who will give us this unified theory?' called one of the other physicists.

'I am close to its completion.' He looked around the tables and smiled inwardly. He had rattled them. It emboldened him to continue. 'I'm not the lone voice you all think I am. Eddington in Britain works on this, too.' A new thought fired him. 'When Johannes Kepler asked Tycho Brahe whether he had observed the parallax and proved the movement of the Earth around the sun, Tycho told him he had not, and said that this proved that the Earth was stationary at the centre of the universe. Kepler believed that all it proved was that Tycho's instruments were incapable of detecting the parallax, and in 1838 science finally developed telescopes capable of seeing the parallax. Without Kepler's belief, he would never have devised his three laws of planetary motion, now thoroughly tested.'

'But this is different,' said Heisenberg, drawing attention back to the front of the room. 'It's not about better technology; it's about there being a fundamental limit to the knowledge we can hold about particles. If we measure the mass of a particle with increasing precision, we automatically lose the precision we have about its position.' As he spoke he regained his earlier certainty. 'The same is true for the energy involved in a reaction and the time the reaction takes. Unless you can show that there is a mathematical error in the derivation of the equation, it shows that uncertainty is woven into reality – that the very act of observing changes the situation.' His eyes filled with courage. 'I will not listen to opinion or belief, even from you. But I will give up the uncertainty principle if you can prove – and I mean prove – that my equation is in error.'

Einstein pursed his lips. 'Very well, I accept the challenge.'

Next morning, he spotted Bohr and Heisenberg chatting over breakfast. He sauntered over, trying not to grin. 'Mind if I join you gentlemen?'

'Please do,' said Bohr.

Einstein pulled up a chair. 'I have a little problem for you both.'

Heisenberg stopped eating.

'It concerns the trade-off between knowing the energy of a process and the duration of the process.' Einstein paused as the waiter poured him coffee. 'Poached eggs, please.'

'Yes, sir.'

Bohr and Heisenberg were as still as statues until he returned his attention to them. 'Imagine a box filled with light and placed on a weighing machine . . .'

They both nodded, knowing that the weight of the radiation would be measured, too, and that this would be an excellent measure of the energy contained within the box.

'Next to the box is a clock. The box opens at a particular time and a single photon is emitted. The weight of the box instantly changes, and that will give us the precise amount of energy released. The clock will have told us exactly when the photon was emitted. Both can therefore be known with certainty. There is no uncertainty between energy and time.'

'It would be impossible to build such an apparatus,' said Heisenberg quickly.

'I'm not saying it would be possible to build such equipment now, but in principle such an experiment is possible. If the uncertainty principle is true, you should be able to give me a theoretical reason why the clock will be affected by the release of the photon. How can those two things be physically related?'

Einstein unfolded his napkin and dropped it on his lap as the waiter served his eggs.

'It's a nonsense without being able to perform the experiment.'

Bohr placed a hand on the arm of his companion. 'He's right. If uncertainty is true, there must be a theoretical way of refuting the experiment.'

Heisenberg looked edgy. 'The successes of quantum theory are huge . . .'

'It takes only a single wrong result to falsify a theory,' Einstein reminded him.

'Enough,' said Bohr sharply. It was unclear whether his irritation was directed at Heisenberg or Einstein. 'There will be a solution. Give me some time.'

'Of course.' Einstein settled to his breakfast. 'These eggs are good,' he said, his grin almost impossible to conceal.

When the day's formal proceedings were brought to a close, the physicists descended on the cloakroom to wrap up for the walk back to the hotel.

Bohr, lost in thought, was making a bad job of putting on his mackintosh.

'Any luck with my problem?' asked Einstein breezily.

'I've not finished thinking yet.'

'But I've heard you discussing it all day with Heisenberg.' Einstein plucked his trilby from the hatstand and flicked his wrist to make it somersault before placing it on his head.

Next day the conference assembled like mourners at a funeral. Conversation was muted as they took their seats in the main room, and the carnival-like atmosphere of the previous days had utterly evaporated.

Bohr stood up. His hair was a mess and he had dark rings under his eyes. 'You have given me a sleepless night,' he said to Einstein, 'but I have an answer for you. When the photon leaves the box, the box will weigh less. It will therefore rise a little on the scales.' Einstein inclined his head in agreement. 'At its new height it will feel a weaker pull of Earth's gravity than before. According to general relativity, the rate at which time passes is dependent upon the strength of the gravitational field. So, time for the box slows down during the emission of the photon. In other words, the emission of the photon causes time to pass at a different rate after the release than before. The difference in this rate introduces an uncertainty into the measurement of the clock standing besides the apparatus.' He passed a sheet of paper down the line to Einstein, who read the tired handwriting incredulously. 'As you see, Albert, the uncertainty predicted by general relativity is exactly the same as predicted by Heisenberg's principle. You forgot your own theory.'

Hot and cold needles lanced Einstein's skin.

'If you'd looked into it closely enough,' said Bohr, 'you'd have seen that the uncertainty principle is buried in there, too. You might even have been able to discover it.'

In an effort to control his embarrassment, Einstein brought a fist crashing down on to the tabletop, setting the pens and water glasses rattling. 'God does not play dice.'

Bohr's voice was contemptuous. 'And when did you start giving God orders?'

Einstein escaped from the conference as soon as he could and tried to lose himself in the walkways of Leopold Park, with its tall trees and open lawns. The damp October air seemed to match his mood perfectly. Occasionally he would stop, halting because all his brainpower was required to sort through the confusion. Things were more perplexing now than ever, worse than before these supposed breakthroughs. Could physics really have reached a fundamental limit in what could be known about Nature?

He and the others would be remembered as the destroyers of physics, when they had thought themselves its greatest champions. Even if there were hidden realities, as he wished to believe, they might as well be discussing angels on a pinhead.

He crunched along the gravel paths, alternating between anger and humiliation until, with a creeping realisation, he sensed he was being followed. He turned and saw a stout priest in a large, unbuttoned coat. The clergyman paused, then lifted an uncertain hand.

Einstein pretended not to notice, turned back and walked on, quickening his pace.

'Professor Einstein?'

Accepting that the only way to escape now would be to run away, Einstein sighed and came to a stop.

'Professor Einstein, I was so hoping to meet you. I hope you don't mind me approaching you like this. I'm . . .'

'You're a priest.'

'Yes, I am. A secular Catholic priest, working as a physicist at Louvain.'

'I'm rather busy, you know.'

The man looked sceptical but remained polite. 'I'd like to talk to you about a dynamic universe.'

'In which case, I'm very busy.' Einstein walked away.

Jogging clearly did not come naturally to the priest, who bounced heavily along the path.

'There's a way of joining your solution to Willem de Sitter's by showing them as start and end points of a cosmic evolution,' he panted.

'The universe is static, everyone knows that, Mister . . .?'

'Lemaître, Georges Lemaître.' He thrust forward his hand. 'The universe expands.'

Pushing his hands into his pockets, Einstein turned to face him. 'You have clearly not kept up with the literature. I introduced the cosmological constant to keep the solutions static.'

Lemaître looked sheepish and Einstein thought he had him. His triumph disappeared as the priest spoke cautiously but firmly. 'Even with the lambda term, your equations are unstable. I've checked them. It's like balancing a pencil on its point; the merest disturbance will break the equilibrium and set it into motion. The cosmological constant cannot help.'

'De Sitter's solution is static as well.'

Lemaître shook his head. 'There's the de Sitter effect that can be interpreted as an expansion. Also, there's a mistake in his mathematics, in the way he chose his coordinates. I've found this too, and once it's corrected, his universe expands.'

Einstein looked through the horn-rimmed glasses to the bright eyes behind. 'You've published this?'

'Yes, sir.' Lemaître drew a deep breath. 'And more. To move from your geometry to de Sitter's, the universe must expand. I have predicted the rate of this expansion.'

'Where is this published?' If Einstein promised to take a look, perhaps he could end the conversation.

'The Annals of the Brussels Scientific Society.'

'How do you expect anyone to read it in there?'

'I'm Belgian.'

'Patriotism, eh?'

'I've also sent a copy to Professor Eddington.'

'Eddington?' Einstein raised an eyebrow.

'Yes, I was his student for a year.'

The eyebrow slunk back into position.

'What does he make of it?'

'I haven't heard yet.'

Einstein cocked his head. 'Perhaps that says it all, Mr Lemaître.'

'But the redshift of the galaxies.'

'The what?'

'The redshift, the systematic movement of the spectral lines towards the longer wavelengths. All the astronomers know about that now. My paper shows that the further away a galaxy is, the larger the redshift will be.'

Einstein scratched his forehead. 'Do you mind if we walk? I must think.'

Lemaître gestured along the path and they set off. The damp leaves clung to their leather shoes. Einstein's mind was full to bursting. How could he think clearly about all of this when his mind was already full of the quantum discussion? Particles you couldn't pin down to a position, and now galaxies that were rushing off into space. It was all movement and madness, when what Einstein really needed was quiet. He needed to be alone.

'Your calculations may be correct,' he conceded as they crossed briefly through an opening in the trees, 'but your physics . . .'

Lemaître looked at him keenly.

'. . . is abominable. It has been established for centuries that the universe is static. Good day, Mr Lemaître.'

The Belgian came to a stunned halt and Einstein walked on.

28

Zurich

1930

Einstein paused to tap out his pipe in the gutter outside, having sparked up only to clear his chest for the march from the station. He slipped the empty briar into his coat pocket and paused briefly, partly to get his breath back and partly to compose himself. In the headlong flight to get here, he had not thought too much about what was waiting for him on the other side of Mileva's door. He clasped his hands together to steady them and went inside.

Mileva's face was all drama. In her relief at seeing him she hugged him tightly.

'I have a seat back on the last train,' he told her. 'I cannot miss it.'

'Just to have you here for the day will be such a help.' She looked so grateful, clearly on the verge of tears.

'Where is he?'

'Where he's been for the last week. In his room, refusing to get up. He hasn't been to his classes.'

'I knew from his letters he was struggling,' said Einstein, 'but I had no idea he was so bad. Thank you for letting me know.'

'I thought his brother leaving home would upset him, but he sailed through that and seemed happy at university.' Her hands had a life of their own, gesticulating randomly or pushing hair away from her forehead. She crossed her arms to trap them. 'But this woman has really made him suffer.

Couldn't she have picked on someone her own age? He was in love with her, completely. Head over heels, silly boy.'

'I'll talk to him.'

She led him to the bedroom where Eduard was a lump under the sheets, facing the wall.

'Tete.' Her son did not move. 'Tete, your father is here.'

The lump rolled over; its eyes widened with surprise, perhaps a hint of shame. Tete pulled himself to a sitting position and hugged his knees. Above the head of his bed was a picture of Sigmund Freud.

'I met Freud a few years ago,' said Einstein. 'We spoke quite amicably, though I think he understood about as much of my physics as I did of his psychiatry.'

Eduard gave him a suspicious look.

'I mean no disrespect. I know you're studying his analysis techniques, and I have to say that, from what little personal experience I have, I am convinced there is something in his methods.' Einstein looked around for a distraction. 'Shall I shut the window?' He made for the open sash, whence the noises of Zurich were spilling into the room. 'It's cold in here.'

'No,' said Eduard, 'I like it open. I feel that I can escape.'

Einstein exchanged a glance with Mileva but let the statement go. He could see that she too thought it best not to acknowledge such melodrama.

Einstein returned and perched on the bed, near his son's feet.

'You know, passions grow and burst like bubbles. It's the way of such things. The first time it happens, it's always difficult.'

'I thought she loved me.' Tete's voice sounded like a child's.

'And she did, for a while.' Einstein looked at Mileva, hugging herself by the doorway. 'Perhaps a glass of water?'

She nodded and left.

Einstein lowered his voice. 'What if you were to find a younger woman? A dalliance would do you good, nothing serious, don't let the heart get involved. It would convince you that there's more life to be lived, more fun to be had.' He nudged his son, whose fleshy arms absorbed the contact like a bag of beans.

'There's no meaning to life except life itself.'

'That's true if you mean there's no higher purpose, no God-given destinies to fulfil, but that doesn't mean you can't fill your own life. Be part of society, find others to share with, discover what's important to you and focus your efforts on it. Then you can live a joyful life.'

Mileva arrived with the drink. Eduard took it without comment and drained the glass. Shortly afterwards he seemed to brighten, but his speech remained stilted and his words were leaden.

'I worry that my letters have been too full of rapture. I fear what you might think of me, Papa. As soon as I post them, I want to snatch them back and tear them up.'

'I treasure your letters for their passion. How could you think otherwise?'

'You're cooler in disposition than me.' He stared straight ahead at the pile of books and writing paper on his chest of drawers. 'It can be difficult to have such an important father. It makes one feel so insignificant.'

'But I'm not important. This hysteria that follows me around is false. It's not who I am to you – it's not who I am at all. The celebrity I've been forced to accept is nothing but an intrusion. I will give it all up tomorrow if it will help you.'

Eduard turned to his father; his eyes were hooded.

Mileva laid a hand on Einstein's shoulders. 'You know that will be impossible, Albert. Don't make promises you cannot keep.' Her voice was deliberately mild. 'They will always hound you.'

Einstein knew she was right; just the rumour of the completion of his paper on a unified theory had brought a horde to his door. He had fled the apartment by the fire escape while Elsa fobbed them off at the door.

'It's absurd,' he whispered, telling them the story. 'Ironic, don't you think, that a man so adverse to authority has become one himself? The reporters turn to me for everything.'

'Did you resist your father?'

Einstein nodded. 'But I don't recommend it.'

The attempt at humour died. Eduard said, 'Psychoanalytically, because you fought your father rather than submit to him, unconsciously you were driven to become an authority yourself in order to replace him.'

'That's what they teach you?'

Mileva's hand tightened in warning on his shoulder at his implied criticism. He changed tack. 'Tell me, Tete, how is your music?'

A few of the clouds on Tete's face cleared. 'I love Chopin.'

'Get dressed and play for me, son.' Einstein held his breath. 'Please?'

Mileva bit her lip.

Eduard thought about the request, and finally nodded.

It was some time later that that the young man shuffled into the living-room, sloppily dressed in a misshapen cardigan, edging his way round the mismatched furniture.

'You share your father's sense of fashion,' said Mileva.

Uncertainty crossed Eduard's face and he hesitated. Einstein smiled to show that it was a joke, drawing an uncertain grimace from his son, who walked on to the piano.

Einstein sprawled on the settee in a pantomime of relaxation, Mileva perched primly on one corner.

Eduard selected a piece of music and placed it on the music stand. Before he played, he placed his hands together in a kind of prayer and bowed his head. His eyes were squeezed shut. When they opened again he was a different person. His arms moved with fluidity, picking out the first chords with simple efficiency to set a melancholy mood.

Einstein recognised the piece as one of the Nocturnes. He guessed at No. 20 in C sharp minor.

Eduard moved into the first of the glissandos, pausing almost too long on the first few of the descending notes but accelerating with such assuredness that his father swallowed in astonishment, more so when Eduard positively attacked the notes on the reprise. The force in Eduard's playing increased, his anger palpable. Einstein saw Mileva wipe away silent tears as they listened to their son's rage.

When he finished the piece the young man looked ready to collapse. He was breathing heavily and his face was red. Einstein willed him to cry, to let everything go. He struggled from the settee and placed a hand on his son's shoulder, but instead of sinking together with him and crying as his father had hoped, Eduard stiffened and checked himself again.

As the light began to fade, so Einstein began to get restless, rubbing his knees in rehearsal for actually having to announce his departure.

Mileva recognised the signs. 'The upstairs room is there if you need it.'

'Thank you, but I have to go.'

'Lectures? Interviews?' She was probing, but gently.

'Margot's wedding,' he said.

'Margot?' The surprise in her voice was genuine.

'I know, we're just as astonished. More so. Her sister was married a few years ago and now lives across the city. Margot and Dimitri will move in with us.'

'She's not moving out?'

'One small step at a time for timid Margot,' said Einstein.

They laughed softly together, the sound filled with understanding rather than mirth. They moved to the hall. Mileva slipped his coat up on to his shoulders and then busied herself with the buttons. He could smell her perfume, blended with the aroma of tobacco that clung to his moustache.

'I do still think about Berlin,' said Mileva.

'It's changing so fast, and not for the better. The people are growing so bitter.'

'I read more and more in the papers about Adolf Hitler.'

'He exploits the hardship of the workforce.' He suppressed a sigh as she finished securing his coat. He looked back into the apartment. 'You think he'll come round?'

Eduard had been calmer during the afternoon but had eventually sloped off to his room.

Mileva said, 'I hope so.'

'I recognise a lot of me in him.'

'Really?'

'Yes, more pronounced though. It's not going to be easy for him, or those around him. Let me know how he is.'

She nodded. 'Thank you for coming today.' There was a painful sincerity in her eyes that made him feel guilty. He was an estranged husband and an absent father. 'I'll always do what I can, Mileva. You know that.'

Impulsively, he kissed her on the cheek; he felt the softness of her skin. Embarrassed, he turned for the door.

29
Louvain

By this time in the evening Lemaître was usually asleep with a mug of cocoa inside him. Having left the observatories of America behind and settled into his lectureship, he had swiftly returned to a more conventional schedule.

Not tonight. He was nocturnal again, still at his desk, the milk-pan unused. Working in the glow of a lamp, he faced the frosty window with his back towards the bed of his single room.

The paper he studied was unmistakable: Hubble had done it.

The American had observed more and more galaxies – pinpointing the distance to each one and discovering that most were far beyond even the Andromeda galaxy. But that was not the most amazing part.

Hubble and his assistants had measured the redshifts and found that the more distant was the galaxy, the larger was its redshift. The paper contained a graph with distance on one axis and redshift on the other. Hubble had been able to draw a straight line through them. It was the perfect confirmation of the de Sitter effect, and exactly the behaviour with distance that Lemaître had predicted in his paper to the Brussels Scientific Society.

There could be no doubt any more: the universe was expanding. All of the galaxies were being driven away from each other by a relentless movement of the void. And Lemaître had predicted it first.

Einstein couldn't ignore him now, he thought.

He read Hubble's paper over and over, shaking his head in astonishment each time he reached the conclusion. His cheeks began to hurt with the effort of smiling, but his joy was soured when he finally turned to the last page and found that his name was not among the references.

A moment of anger lanced through him for sending his work to the Belgian journal. He should have published in English or German. Perhaps then the paper would have received more attention, and it was puzzling that Eddington had not written back to comment.

Lemaître knew that he needed to be bolder about his ideas. That was no sin. After all, what was the good of doing the work if it wasn't being read?

He trembled as a thought struck him. It was an idea, really, one that he recognised had been trying to grow inside for a while. He had hardly dared acknowledge it before, but now that the expanding universe had been proven he realised he could nurture the truly astonishing idea that lay within.

Epiphany swept through him.

This time, when he was finished, he would let people know. He would target Einstein with personal letters. As for the rest of the scientific community, he knew exactly where to start: at Harvard, where he had been invited to a meeting of the International Astronomical Union early the following year.

That should give him enough time to perform the calculation. He lit a cigarette to calm his mind and pulled a fresh sheet of writing-paper on to the blotter. He would not be sleeping tonight.

30

Berlin

On the day of Margot's wedding there was a Nazi parade. Many of the marchers were dressed in quasi-militaristic, brown-shirted uniforms with armbands that bore jagged swastikas. Einstein felt close to panic as he watched them pass the Berlin marriage bureau where the guests were gathered.

It was not the same as the joyous, if misguided, march to war Einstein had witnessed when he first arrived in Berlin. This parade was regimented and dangerous, oozing power and aggression.

'You'd think they were already in power,' said Walther Nernst, trussed up in his best suit next to him.

'I fear it won't be long. They increase their share of the vote at every election.'

Nernst peered out through cloudy eyes. 'Then what's to become of us? I'm sixty-six, my career's on borrowed time, and they're not going to want a Jew in a position that can be filled by a Nazi half my age, with half my knowledge.'

The march progressed relentlessly, forcing people to scurry out of their way. From the sidelines onlookers watched, a mixture of curiosity and uncertainty on their faces.

'It's no use talking to Max. He thinks we should just pretend nothing's happening.'

Einstein nodded sadly. 'I used to trust him. I used to look up to him. Really look up to him.' It was as if nothing was as it once had seemed. 'How can we pretend that nothing is happening?'

'I know. Try telling Fritz.'

'Haber? What do you mean?'

'My God, haven't you heard? He's gone already. The Nazis have hounded him out. He's heading for England.' Nernst rolled his moustache between his fingers. 'Still, today's not the day for worries, eh, Albert? They're a good-looking couple, Margot and Dimitri. Where are they honeymooning?'

Margot was dressed in green velvet with grey fur trim. Her husband stood proudly next to her, both of them chatting easily with the guests.

'We're leaving them in the apartment. It's Elsa and I who are clearing off, back to America. Pure science this time: two months' research at Caltech, no fundraising, no worries, no politics – just science.'

Inevitably word had spread of his visit, and new offers of appearances were being telegrammed through every day. Thankfully Elsa was dealing with all of that, flatly refusing the lot no matter the sums of money involved. As she rightly said, playing hard to get only increased the fees.

Nernst was studying him. 'It's interesting the way you say that.'

'What?'

'America. You say it with affection.'

The sound of Margot taking in a sharp breath caught Einstein's attention. Her earlier smiles had been replaced with a look of incredulity. Einstein followed her gaze across the street and gasped too.

The Nazis had gone, and in their place was a small dark-haired woman in a shabby overcoat. She looked bedraggled and desperate.

Mileva.

Einstein was already in motion, rushing down the bureau steps and across the road. Mileva was swaying like a stalk of wheat in a breeze, and now that he was with her she lost all cohesion. She dissolved in his arms, her body convulsing with sobs and her mouth emitting an awful wailing.

A black seed took root inside him. 'It's Tete, isn't it?'

She nodded into his shoulder.

'Is he . . .?' His voice failed.

Through her anguish, she uttered, 'No, he's still alive.'

'Thank God.' The relief gave Einstein some clarity. He soothed her in his arms until the sobbing subsided. 'You must have travelled through the night. Let me take you for something to drink.'

With a furtive look over his shoulder he saw Elsa and the others in the wedding party watching him, aghast. These two parts of his life should not overlap. He waved uncertainly and led Mileva away.

He chose the first café they came to, opening the door for her. She was still dabbing tears from her eyes. What an odd couple they must have looked as they sat down, Einstein in his suit and striped tie and Mileva bedraggled in her tatty coat.

The blond-haired waiter took one look at the pair and turned to serve another couple. 'You can wait, Jew.'

Mileva looked shocked.

'It's like this all the time,' said Einstein, trying to brush it aside. 'Tell me what happened with my boy.'

Her eyes filled again but she controlled it. She fixed her gaze on the salt and pepper pots at the centre of the table.

'He tried to throw himself out of his window when you left. It was all I could do to hold on to him. He's so big now, so heavy. I scratched him deeply. I didn't mean to. I was desperate, clawing at him to try to hold on to him. I drew blood.' She pulled her fingers down her face, across her jaw.

'Where is he?'

'In the hospital. Sedated.'

'Does Albert know?'

'I've written to him.'

Einstein's mind was in uproar. 'But Tete seemed so much better when I left.'

Annoyance flashed in her eyes.

'Sorry,' he said, 'It's difficult for me.' That was a stupid thing to say. He rubbed his brow, searching the room for some better inspiration. He noticed a watermark on the ceiling in the corner that had been painted over to try to mask the damage. 'Oh, my boy, what have you done?' he muttered over and over, a mantra to calm himself and get him thinking again.

The happy chatter of the other customers buffeted him. He caught snatches of discussions about last night's theatre trip and the price of petrol, all of it detestable bourgeois nonsense.

'I can't do this alone, Albert.'

Don't you think I know that? he wanted to snap at her.

'I need your help. I need you to . . .'

'I'm going to America tomorrow,' he blurted.

He might as well have slapped her in the face.

'America?'

'Yes.'

'Why didn't you tell me the other night? You said you'd do anything.'

'Anything I could. I meant it.'

'But how can you, if you're going to America?' Her voice drew the attention of the waiter, busy holding a chair for a pencil-slim woman with immaculately-styled hair.

Einstein chewed his lip. Gears were shifting inside him, grinding his insides together.

'That's always been your problem, hasn't it, Albert? You're great at the general principles, but the specifics escape you completely.' There were no tears in her eyes now. 'You say that you'll help, but you can't muster the least specific action. You used to say you loved me, yet you couldn't do anything to nurture our marriage.'

'That's unfair, Mileva.' He should have been getting angry with her, but instead he was feeling a growing panic.

'Is it really unfair, Albert? Really?'

It took all his willpower to stay in the chair. He wanted to run. But where to? Back to the wedding? Swap this family duty for that one? The wall he had carefully constructed between the two families was crumbling away. Mileva's hard stare was cutting into him. They had shared so much, produced a family. Margot and Ilse, yes, he loved them – in a way – but not like the boys. Blood was blood, and those ties go to the very bone. He knew that now.

Tete needed his father. Mileva needed her husband.

But Elsa was so loyal, so undemanding in a way. They had shared the experiences of seeing the world, constant companions in a way that he and Mileva had never been.

He couldn't live without Elsa. She took the burden of practical responsibility off him; she made his life easier – not that it seemed at all easy right now.

Mileva's voice cut into his thoughts. 'If you really wanted to help, you would stay. Nothing else would matter.'

He was numb. Elsa's face, horrified as he led Mileva away from the marriage bureau, floated into his mind, blotting out the real image of his first wife.

He watched, paralysed, as Mileva stood up. 'Goodbye, Albert. It may take you a while to understand this, but I'm glad I came today. I'll write to you.'

He tried to do something. Anything. He imagined standing and catching her arm. *What could be more important than Tete?* he would say, and they would travel back together. That was the right thing to do, surely?

But by the time he realised that this was his intention, she was almost at the door. There was a determination in her movements that he had never seen before, and this gave him more pause for thought. She pulled open the door, the breeze catching her already untidy hair, and she slipped into the street. A moment later she was lost to the crowds.

Einstein waited a long time, struggling to understand his own inertia. At last he heaved himself to his feet. He had made the decision, or rather it had been made for him.

The surly waiter squared his shoulders. 'And shut the door behind you.'

31

Harvard

1931

It was hard not to catch the astronomers' excitement. The conference attendees were gathered on the staircases and in side rooms; they halted in their tracks and blocked entrances in their enthusiasm for discussion. Cosmic secrets that had been concealed for millennia were falling open so fast it was becoming impossible to keep up with it all. Astronomers were dividing into specialities, stars and galaxies, and gatherings like this one had become essential for them to learn about each other's work.

Lemaître spotted Eddington almost at once. His former mentor was navigating the assembled astronomers with the kind of old-fashioned courtesy that meant he could acknowledge everyone without ever inviting them to detain him. They all wanted to talk to him, of course. He was the man who had proved relativity, yet he was as self-contained as ever.

Lemaître had learned at Cambridge that the only way to halt Eddington in public was to ambush him, so he hid behind a group busily discussing spectral lines in red-giant stars. When his quarry was so close that Lemaître could smell the stale tobacco, he stepped out into his path.

'Hello, old chap,' said Eddington reflexively, coming to an abrupt stop. His expression was essentially the same, although deeper jowls and deeper creases in his brow had shifted it from smug to somewhat perplexed. 'I was hoping to see you.'

'Really?' said Lemaître.

'What's Einstein up to? All this nonsense about hidden realities and refusing to use the quantum theory . . .'

'Don't ask me; I'm as confused as you. He and I generally stick to relativity.' After Hubble's discovery Einstein had suddenly started taking more interest in Lemaître's work. 'He's in Caltech working on his unified theory, now' said the priest. 'He's invited me over there after this meeting to give some lectures, but I'm still concerned. Has he always been so stubborn?'

'Oh yes,' said Eddington, winding up for his punchline. 'The only difference is that before, he was right.'

'Do you think a unified theory is possible?'

'It's possible – got some ideas myself – but Einstein hasn't found it. Take that last paper of his.' Eddington's expression became wry. 'Do you know they exhibited a copy in the window of Harrod's? It pulled a crowd so large it nearly stopped the traffic. Not that they could understand a word of it, of course. But then, I'm not sure I could either.'

'Me neither.' Lemaître had felt empty reading it. Einstein was chasing shadows, drowning in preconceived beliefs instead of letting the universe guide him.

'Aha! Just the two I've been waiting to see.' The small figure of Harlow Shapley jostled in between them. 'What do you two make of all these observations coming out of Mount Wilson? What's Hubble up to?'

'The observations make perfect sense,' said Lemaître firmly, 'The universe is expanding.'

Shapley exhaled noisily. 'He'll be insufferable now. What I want to know is, what the hell is it expanding into?'

'Space itself is expanding,' said Lemaître. 'The galaxies sit in space, being carried ever further apart like currants in a rising dough mixture.'

'Trust you to think of a culinary analogy,' quipped Eddington.

Lemaître ignored him. 'The universe needs nothing to expand into because there is nothing outside the universe. It is everything.'

'And how the hell do we explain being at the centre of the expansion?' Shapley said in his excitable drawl.

'It's an illusion,' said Lemaître. 'We're not aware of our own motion; instead we see it imprinted on the movement of the other galaxies. If we were to live in one of the distant galaxies, we would see the Milky Way flying. That's just relative motion.'

'Relativity again, huh? Give me the stars any day,' said Shapley. 'Still, at least you both seem to understand it.'

Eddington wagged a finger. 'But that's just the problem. We can't, Einstein can't. His solution was a static universe, so was de Sitter's. Seems as if general relativity can't explain the universe after all. The question is: how do we reconcile it? Shall we put a little motion into Einstein's universe, or a little matter into de Sitter's?'

Lemaître stared in disbelief.

'What?' said his former mentor. 'Have I just committed a blasphemy?'

'I sent you the solution. Do you not remember my paper? I sent it at the time of Solvay.'

Eddington's face fell. 'Oh my, I'd forgotten. You've had a crack at this, haven't you?'

'A crack? I solved it. I showed how Einstein's universe could evolve into de Sitter's by expanding. I predicted the rate of expansion and got an average of 625 kilometres per second. Hubble's value is around 500. I was so close.'

'You predicted that?' asked Shapley.

'I did.'

'And you believed it?'

'Of course.'

Shapley whistled.

Eddington looked stricken. 'Sorry, Georges, I've let you down badly. I'll arrange publication in *Monthly Notices* as soon as I get back.' He sounded genuinely contrite.

Lemaître had never witnessed that before. He suddenly felt resigned. 'Probably no need to include my prediction of the expansion rate now, is there? Now that Hubble's measured it, I mean.'

'As you wish. I suppose it would seem like fixing the stable door after the horse has bolted. Does Einstein know of your solution?'

'Oh yes, that's why he invited me. The trouble is, I've changed some of my thinking. Originally I thought his closed universe could be the starting point, but it can't; the calculations show that it's unstable. I can find nothing to limit the original size of the universe.'

'But how small would it get?'

'That's the thing. I can't see anything to stop it until it runs into an infinitely dense point, and if that's not bad enough, before it gets to that point it's going to get so densely packed that the laws of quantum theory are going to take over.'

Eddington whistled. 'And you're going to tell Einstein all of that?'

'I am,' said Lemaître.

'Then you're a braver man than I.'

32
Pasadena, California

When they greeted each other on the steps, Einstein's face looked greyer than Lemaître remembered. It was as if the ash from his pipe had invaded his flesh. Although the physicist smiled, it was a mechanical movement of his mouth. His eyes took no part in his expression.

'Welcome to Caltech,' he said flatly. His English was much improved, though still spoken with a heavy accent.

It was early January and the air was not unpleasantly cold. Louvain would have to wait until spring for such a temperature.

They turned to climb the stairs and were accosted by a man in a raincoat and a pork-pie hat. 'Mister Einstein, can you explain relativity for our readers in a sentence?' The reporter stood poised with his pencil and notepad, as though taking a wicket-keeper's stance.

Einstein's face flashed into irritation 'It would take me three days to give you even a short explanation.'

'Well, what about the fourth dimension? Can you define the fourth dimension in a word?'

'Ask a spiritualist ' He stepped around the reporter and was just opening the door for Lemaître when the reporter called again, 'What do you think of Adolf Hitler?'

Einstein's head turned. 'He's living off the empty stomach of Germany. As soon as economic conditions improve he will no longer be important.'

The reporter's eyes widened and he scribbled down the quote. He touched his hat and sprinted off.

Inside the airy stone corridors there was a quiet buzz of conversation, as in a factory where one could hear the distant whirr of manufacturing machines, and the place smelled as fresh as the outdoors.

'Do you honestly believe that about Hitler?' asked Lemaître.

Einstein's brows knitted together. 'What else can I believe? Germany is becoming more dangerous by the day, and not just to us Hebrews but to everyone.'

Einstein did not have to open another door during their passage; students and staff would pause, waiting for him to step through when they saw him coming. Lemaître could not help but notice their reverence; they all but lowered their gaze as he passed.

'I read an interview with your wife in *Time* on the train journey,' he said by way of small-talk.

Einstein gave a small laugh. 'Oh, that.'

'Forgive my question, but does it bother you that they report the irony of your not being able to understand your household accounts? It seems intrusive.'

'They are bizarrely fascinated with me; they're like children.' Einstein paused by an office with his name on it. 'But I think I quite like it. And in answer to your question: no, it doesn't bother me, because it's true. I couldn't tell you the price of a cauliflower.'

The office was large, with a wide window looking out on to the tree-clad mountains. Lemaître fancied it was the air rolling down from those peaks that filled this place with its fresh smell.

'That's a wonderful view,' he said.

Einstein glanced at it. 'Yes, but I still prefer Princeton. Not quite so dramatic, not quite so . . . full of itself.'

They settled themselves in easy-chairs.

'Seems I owe you an apology, Georges. The universe is expanding after all.' A look of genuine puzzlement crossed his face. 'Who could have predicted it? Except you, of course.'

'The expansion is not the most amazing part.' Lemaître's heart accelerated. 'It leads to another hypothesis. Something more that I would like to talk to you about.' Einstein nodded cagily for him to continue. 'If the universe is expanding, carrying apart all the other galaxies, then it stands to reason that everything must have been much closer together in the past. If you go back far enough, all the stars will have been as crowded together as the automobiles I saw in Times Square.'

Einstein's face was blank. Deciding this was not the worst reaction he could have expected, Lemaître pressed on. 'Go back further, and all the matter in the universe must have been compacted together into a single mass that exploded and set the expansion in motion. To give the stars and galaxies enough time to form, I calculate that the universe must be at least ten thousand million years old.

'The universe is infinitely old.'

'It can't be. Hubble's expansion proves that. Your equations prove that. Everything must once have been compacted together into some kind of primeval atom that split apart and led to the beginning of space and time, a moment at which the evolution of the cosmos began.'

Clouds of suspicion had gathered on Einstein's face. 'You're talking about the Creation?'

'No, no, anything but that.'

'Yes, you are. The beginning of space and time. Genesis. You're trying to force your religious beliefs into science.'

Lemaître met Einstein's accusatory gaze. 'Never. The conclusion is to be found in your own mathematics.'

'One must never confuse science with religion. The Old One is to be found . . .'

'The Old One?' Lemaître lifted an eyebrow.

'You of all people should be familiar with him.'

'So you're not an atheist then, after all?'

'Of course not,' Einstein said as if it were obvious. 'My problem is with religion, not with God. I believe in a God who is Nature, and vice-versa.'

Lemaître deliberately softened his voice. 'We have more in common than you think.'

'No! I will have nothing to do with your God. He is too personal for my taste.'

'Personal?'

'Too full of petty human emotions to be anything other than a construction of man's.'

'How can you say that? The Trinity is there to explain both personal and impersonal aspects of God. You are prejudiced. Perhaps it is you that confuses science with religion. Science is not a route to salvation. Remember what Galileo is reputed to have said: the Bible tells us how to go to Heaven, not how Heaven goes.'

'And look what your lot did to him.' Einstein jabbed a finger towards the flash of white at Lemaître's throat.

Though the insult cut, Lemaître let it pass. 'If we reverse that sentiment, then we see that general relativity is not essential for our salvation. If it had been, it would have been revealed truth and described in the Gospels. It isn't, yet we still pursue such knowledge because it comforts and informs us. It's entirely different from religion. I would never mix the two. They are quite separate.'

'Religion means nothing to me, in fact worse than nothing. On my voyage back from Japan I was persuaded to stop in Palestine and see the construction taking place for the new Jewish state.'

Lemaître held himself rigid, wishing to betray nothing. He had come here to discuss science, not religion.

'I visited the Holy Wall,' continued Einstein, 'and I looked on in horror at the worshippers, wailing and weeping.' He made an harrumphing noise in the back of his throat. 'Complete loss of dignity. Religion is nothing but the sum of human weakness, if you ask me.' Einstein stood up, pointing angrily. 'And now you come here, trying to . . . to . . . infect science with Genesis. False! A scientist should never work to belief. It cripples his objectivity and leaves a poor taste in the mouth.' He took out his pipe and began patting down his pockets for tobacco.

'Why persist with your unified theory?' asked Lemaître.

'Because Nature loves the beauty of simplicity.'

'How do you know that?'

'I feel it.'

'A belief? In a scientist?'

Einstein shook his finger. 'An intuition, Georges. Nothing to do with religion.'

'And neither is the primeval atom.'

'Not that again!' Einstein waved dismissively, growing more agitated in his search for tobacco.

Lemaître was determined now. 'Think about radioactivity. On average, a radium atom sits around for seventeen hundred years before it splits apart. What if everything was contained in a single primeval atom that split apart and began the universe? No religion there.'

'But if it's an atom, that would mean the beginning of the universe was governed by quantum laws.'

'Then so be it. It's the place where relativity and quantum theory collide. Any unified theory must be capable of addressing the beginning of space and time. We are pursuing the same end, just along different paths. My agenda is not religious. It comes from your mathematics. Nothing else. It means that once, in the distant past, there was not the Biblical creation but a beginning.' Blood pounded in Lemaître's veins. He thought furiously, desperate for the words that could encapsulate his insight and placate Einstein at the same time. On the verge of despair, he found the words he had been searching for. 'There was once a day without yesterday.'

The room filled with silence. Einstein was motionless, empty pipe in hand, eyes fixed on a point on his desk. It took him several moments to realise that the conversation had faltered.

Lemaître's chest heaved; his host had not even been listening.

'I'm sorry,' said Einstein, 'I had some bad news this morning.'

'Anything you would like to talk about?'

'No,' said Einstein quickly, then added, 'my younger son has been diagnosed as having schizophrenia.'

Lemaître bowed his head. 'I'm truly sorry to hear that. Perhaps we should postpone our discussion to another day.'

'No, no.' Einstein gestured for him to remain seated. 'Listen to me. A Jew confessing to a Catholic priest.' He began searching his desk again and pulled out an empty tobacco pouch from under the papers. 'My wife rations me. Says it's making me cough.'

Lemaître fumbled in his pocket and passed over some cigarettes. 'Have one of these.'

A boyish giggle escaped Einstein. It collapsed the moment after it was born. 'I don't laugh as much as I used to.'

'It's like riding a bike. You never forget how.'

'I used to tell my boys that life was like riding a bike, you had to keep going forward in order to retain your balance.'

'I used to ride, years ago. I liked to stop every now and again and look around. The more we look, the more we discover.'

Einstein was splitting open the cigarettes and stuffing the contents into his pipe. 'Do you think we will ever understand it all?' he asked.

'There's only one being who knows the answer to that.'

Einstein's lips moulded themselves into a crooked smile. 'Well, it certainly isn't Heisenberg.' He lit the pipe and leaned back in his chair. 'Tell me again about your primeval atom. I would like to understand your ideas before your lecture.'

The car dropped Einstein at his hotel. His thoughts were so full when he arrived at the top floor that he failed at first to notice the open packing-cases strewn about the suite. He was shaken out of his reverie by Elsa's shrill voice ordering the maid about.

'What on earth is going on?' he roared.

His wife came running from the bedroom. 'Thank God you're home. We have to go, we have to get back to Berlin.'

'Why?'

'Margot. The apartment's been broken into.'

'What?'

'Raided. Margot was there, alone. Can you imagine? She must be a nervous wreck. Says they came back a second time and searched again.' Elsa was throwing things into the cases without any attempt at sorting them. 'We must go home.'

'Elsa, wait. If they're raiding the apartment, they're looking for a reason to arrest me. We can't go running into their arms.'

'But Margot . . .'

'Dimitri will look after her.'

'He wasn't there. He was out. That's another story. There was a riot. He was caught in it, nearly hurt.'

'Nazis again?'

'Who else?' she spat.

'Elsa, please think about this.' He tugged a chair free from under the table and sat down heavily. 'Firstly, how do you know all this?'

Elsa showed him the telegram. He picked it up and read the terse message. Instead of the dread he expected, it brought about a great sense of calm. He now understood the phrase 'cold rationality'. All his science had been conducted with rationality, yes, but there was nothing cold about it when he calculated. It was laced with the bright stuff of passion, but now he sensed clarity, even destiny. The feeling worked through him, into his very bones.

'Come and sit down, Elsa.'

She seemed to sense it too. She stepped over the packing and sat at the table, close to him.

He put his hand on hers. 'Our time in Germany has come to an end.' She stared at him blankly. 'We've both known this has been coming, ever since Rathenau. We just haven't been able to decide where or when. Now the decision is made for us. It's time to accept one of the American offers.'

'We can't leave the girls.'

'They're adults, and married. They can choose to come with us or not. But it is time for us to leave. Berlin is no longer the same as it was, and it's only going to get worse.'

She nodded dumbly, and they sat for what seemed like an age. Her eyes were full of tears.

'Elsa, I've been married twice, and neither time have I made a very good job of being a husband. There's nothing I can do about my first marriage, and, to be honest, worrying about it has probably not helped me concentrate on my second. But, if we start again on this side of the Atlantic, at least I'll no longer have that distraction. Perhaps I can begin to make amends with you.'

The maid appeared with a pile of clothing. She stopped in her tracks when she saw the scene at the table.

Elsa's cheeks were streaked with tears. Einstein tried to smile at her. 'Tell me, Elsa. Should I telegram Princeton?'

She sniffed loudly and produced a handkerchief. Dabbing her eyes, she turned to the maid. 'Take those back to the wardrobe, please. We're staying.'

Einstein could see Lemaître pacing nervously outside the lecture hall, reciting key parts of his presentation. As he approached he said, 'So, a day without yesterday, you say?' Lemaître nodded cautiously. 'Even if I were to accept that your beginning of the universe is to be found in my theory, how could we prove it?'

'I've been reading about balloon experiments that show increasing amounts of radiation at high altitude. If the universe began in fireworks, I think these cosmic rays may be the left-over sparks. And that makes all the stars and galaxies the smoke remaining from that great beginning.'

Einstein puffed out his cheeks. 'You don't think in small measures, do you?'

'Neither do you.' Lemaître felt a little bashful. 'Your unified theory.'

Einstein's eyes were suddenly alight. 'You pursue your primeval atom; I'll search for my unified theory. And let us hope that at least one of us finds the Promised Land.'

'Agreed.'

Einstein stepped away from the door. 'Ready?'

'Aren't you coming, too?'

'I'll slip in at the back. I've arranged for the press to come and listen to you. They'll make sure your ideas are widely heard this time, but if I come in with you, they'll only be interested in me. So, you're on your own. Tell them about the beginning of the universe. I want some peace and quiet. Now, in you go.' He opened the door to a wall of men with cameras and notebooks. 'They want to hear about the day without yesterday.'

Lemaître stepped through and the first flashbulb went off like a star. The world was about to change.

Epilogue

Georges Lemaître's idea of a beginning to space and time created a wave of scientific and public interest. However, his hypothesis that the cosmic rays were the left-over embers of a beginning did not stand up to scientific scrutiny, and academic opinion turned against this idea.

In 1936 Lemaître became a member of the Pontifical Academy of Science, later rising to its presidency. In 1951, however, following a meeting of the Academy in Rome, Pope Pius XII publicly inferred that Lemaître's ideas could be equated to the Biblical Genesis. The claim made headlines around the world.

At a private audience with the Pope a few months later Lemaître explained his position on keeping science and religion separate. This seems to have been the catalyst for decades of internal discussions at the Vatican that culminated in 1987 when Pope John Paul II issued a letter to celebrate the 300th anniversary of Newton's work on gravity. It stated: *science can purify religion from error and superstition; religion can purify science from idolatry and false absolutes. Each can draw the other into a wider world, a world in which both can flourish.*

Nevertheless, back in the 1950s, for Lemaître the damage had been done. To many it confirmed their suspicion that he was in fact trying to foist religion onto science.

In 1966, as he lay in hospital suffering from leukaemia, word reached him of an incredible discovery. Two radio engineers had serendipitously discovered that the universe was bathed in a glow of microwave energy. These microwaves outnumbered atoms by a staggering billion to one and carried so much energy that they could only have been created at the beginning of space and time. It was the proof that Lemaître had been right after all, that space and time had a beginning. We now call this beginning the Big Bang.

A few days after receiving his vindication Lemaître passed away.

In 1933 Albert and Elsa Einstein moved to America to begin the formal process of obtaining permanent residency. Upon hearing that the couple could no longer return to Germany, Mileva offered them refuge in Zurich. They declined and settled in Princeton, New Jersey. Elsa died just three years later from heart disease.

Although Einstein came to Europe often, his visits to Eduard dwindled to nothing. His correspondence with Mileva remained cordial, occasionally warm, as she devoted herself to Eduard's care until her death in 1948. By this time Eduard was residing in the Burghölzli sanatorium in Zurich, where he spent the rest of his life.

Einstein maintained a relationship with Hans Albert, especially after his elder son moved permanently to America in 1938 and built a successful career in hydraulics. Married with three children, Hans Albert travelled the world to supervise construction projects.

Einstein never accepted the quantum theory and worked on its possible replacement until his final day. He died in the early hours of 18 April 1955, having spent the previous evening with Hans Albert, talking and performing calculations.

To this day no one knows how to reconcile general relativity with quantum theory. They are fundamentally incompatible. Trying to resolve this dichotomy lies at the heart of modern physics.

Acknowledgements

First things first: a special thank-you to Nikki, my wife, for seeing this trilogy through with me. This book is dedicated to her.

As with the previous two books in this series, bringing the characters in this novel to life would have been impossible without the exemplary work of previous biographers, translators and historians. Two particularly rich volumes are *Einstein: His Life and Universe* by Walter Isaacson and *The Day Without Yesterday: Lemaître, Einstein and the birth of modern cosmology* by John Farrell. As you can see, both John and I were equally struck by Lemaître's most famous phrase.

Other sources of mine include: *The Day We Found the Universe* by Marcia Bartusiak, *E=mc²* by David Bodanis, *The Born–Einstein Letters 1916–1955* by Max Born, *Tycho Brahe's Path to God* by Max Brod, *Afterglow of Creation* by Marcus Chown, *Edwin Hubble: Mariner of the Nebulae* by G E Christanson, *Einstein's Jury: The Race to Test Relativity* by Jeffrey Crelinsten, *Einstein's Nobel Prize: A Glimpse behind Closed Doors* by Aant Elzinga, *Cecilia Payne-Gaposchkin: An Autobiography and Other Recollections* (Katherine Haramundanis, ed.), *Einstein in Berlin* by Thomas Levenson, *Einstein's Mistakes: The Human Failings of a Genius* by Hans C Ohanian, *Edwin Hubble, the Discoverer of the Big Bang Universe* by Alexander Sharov and Igor Novikov, *Big Bang: The Most Important Scientific Discovery of all Time and Why*

You Need to Know about It by Simon Singh and also www. firstworldwar.com.

And, of course, the writings of the men themselves: *Relativity: The Special and General Theory* by Albert Einstein, *The Primeval Atom: An Essay on Cosmogony* by Georges Lemaître.

This is the final book in the *Sky's Dark Labyrinth* trilogy. As with the previous two volumes I have distilled certain events in the hope of serving the story better. In doing this I have learnt that both art and science are about reducing a plethora of specific experience, observations or details into a general truth that can be shared. I hope that in my presentation of these characters and events I have found some of their truths.

In the narrative, I have used my characters' estimates for distances to the celestial objects and the age of the Universe. The modern values are that the Andromeda galaxy is roughly 2.5 million light years away, and the age of the Universe is approximately 13.7 billion years.

This time around I did not have to invent a fictional character to help fill in the blanks that are missing from the stories, and I was helped immeasurably by photographs of the protagonists. Seeing the emotions caught on their faces provided powerful inspiration.

I have lived and worked with this trilogy for the better part of a decade now, and there seems to be no slowing down in people's interest in the project. Thanks to everyone who has shown support and understood the motivation for writing this story in fictional form.

Science is a human endeavour, and so is storytelling. Let us continue to tell stories about science to entertain and inspire.

Thanks in particular go to: Peter Tallack, Hugh Andrew, Neville Moir, Maria White, Alison Rae, Jan Rutherford,

Kenny Redpath, James Hutcheson, Sarah Morrison, Vikki Reilly, Anna Renz, Edward Crossan, Duran Kim, Hamish Macaskill, Anna Rantanen, Kim McArthur, Ruth Seeley and Nic Cheetam.